SCIENCE INTERRUPTED

EXPERTISE

**CULTURES AND
TECHNOLOGIES
OF KNOWLEDGE**

EDITED BY DOMINIC BOYER

A list of titles in this series is available at cornellpress.cornell.edu

SCIENCE INTERRUPTED

Rethinking Research Practice
with Bureaucracy, Agroforestry,
and Ethnography

Timothy G. McLellan

CORNELL UNIVERSITY PRESS   **ITHACA AND LONDON**

First published 2024 by Cornell University Press

Library of Congress Cataloging-in-Publication Data

Names: McLellan, Timothy G., 1987– author.
Title: Science interrupted : rethinking research practice with bureaucracy, agroforestry, and ethnography / Timothy G. McLellan.
Description: Ithaca : Cornell University Press, 2024. | Series: Expertise : cultures and technologies of knowledge | Includes bibliographical references and index.
Identifiers: LCCN 2023020290 (print) | LCCN 2023020291 (ebook) | ISBN 9781501773327 (hardcover) | ISBN 9781501773334 (paperback) | ISBN 9781501773341 (epub) | ISBN 9781501773358 (pdf)
Subjects: LCSH: Agroforestry—Research—China, Southwest—Evaluation. | Agroforestry—Research—Documentation—China, Southwest. | Research institutes—China, Southwest—Evaluation. | Bureaucracy—China, Southwest.
Classification: LCC S494.5.A45 M44 2024 (print) | LCC S494.5.A45 (ebook) | DDC 634.9/9091648—dc23/eng/20230714
LC record available at https://lccn.loc.gov/2023020290
LC ebook record available at https://lccn.loc.gov/2023020291

Contents

Acknowledgments

I owe enormous personal and intellectual debts to colleagues at my ethnographic site in southwest China. Thank you for being such wonderful hosts during my time in China and for the generosity you have continued to show me since.

My thanks to my graduate school classmates Laura Cocora, Perri Gerard-Little, Scott Sorrell, and Namgyal Tsepak. I cannot imagine having completed this research without your friendship and support. I also owe a great deal to fellow Cornell anthropology grads Can Dalyan, Nidhi Mahajan, and Mariana Saavedra-Espinosa. Thank you for your vital last-minute comments on this manuscript and, more importantly, for a decade of friendship and mentorship. During my time at Cornell, I also benefited from the guidance and support of a much wider community of students and faculty. My thanks to Toby Goldbach, Darragh Hare, Emily Levitt, Vincent Ialenti, Hirokazu Miyazaki, Paul Nadasdy, Lucinda Ramberg, Annelise Riles, Steven Sangren, Emiko Stock, Yu Xingzhong, and the many others who taught me so much at Cornell.

During the penultimate year of my PhD (2016–17), I had the privilege of being an unofficial guest in the Department of Social Anthropology at the University of Manchester. Postgraduate seminars and Sandbar gatherings were an invaluable testing ground for early versions of the ideas and analysis I present in this book. My thanks to Ben Eyre, Maia Green, Soumhya Venkatesan, Chika Watanabe, and all the students and faculty who made my time in Manchester so productive.

From Cornell, I joined UC Berkeley as a postdoctoral fellow in the Center for Chinese Studies. Berkeley was a wonderful environment in which to write the first drafts of this book and provided the opportunity to flesh my ideas out in dialogue with faculty, students, and postdocs across the Bay Area. For conversations about this project, I thank Lisa Brooks, Jennifer Hsieh, You-tien Hsing, Peiting Li, Adam Liebman, Stanley Lubman, Yan Long, Juliet Lu, Jesse Rodenbiker, Tobias Smith, and Winnie Wong. I am also grateful for feedback from audiences at presentations hosted by UC Berkeley's Center for Chinese Studies and UC Santa Cruz's Department of Anthropology as well as to fellow participants in the UC Berkeley History of Science/Science and Technology Studies Writing Group and the Cal science and technology studies retreat.

While I was at Berkeley, the Center for Chinese Studies funded a manuscript workshop for this book. That workshop was vital for clarifying and refining this project on so many levels. As importantly, the enormous generosity of the workshop participants reinvigorated my faith and enjoyment in this project. My deepest thanks to Don Brenneis, Todd Sanders, Kristin Sangren, Rachel Stern, Sarah Vaughn, and Li Zhang. It was such a privilege and such fun to spend a day discussing this book with you.

From Berkeley, I joined the staff at Northwestern University's Institute for Global Affairs. I took a break from writing while at Northwestern, but during my time there, I learned an enormous amount about contemporary universities, what is wrong with them, and the inspirational people who struggle against the odds to make them better places. This book doesn't scratch the surface of what colleagues at Northwestern taught me, but they have nevertheless shaped the pages that follow in important ways. Special thanks to Diego Arispe-Bazán, Patrick Eccles, Liz Jackson, Julie Petrie, and Ariel Schwartz.

By the time I resumed work on this book, I had joined the Bachelor of Arts in Language and Culture program in the Faculty of Arts at Chulalongkorn University. I am grateful to my colleagues here for helping me navigate bureaucracies every bit as troublesome as the ones I describe in this book and to my students for reminding me of the educational values that are at stake in transforming universities.

At Cornell University Press, I thank Jim Lance and Dominic Boyer for supporting the project, as well as the wider team at the press and at Westchester Publishing Services for their work to create this book. My thanks as well to two anonymous readers whose comments helped me make substantial improvements.

My thanks to those who financed this research. Field research for this book was funded by a National Science Foundation Doctoral Dissertation Improvement Grant, No. 1357194. In addition, I received several grants from Cornell East Asia Program: a Lam Family Travel Grant, a C.V. Starr Fellowship, and a Lee Ting-Hui Fellowship. Further support was provided by a Cornell Society for the Humanities Sustainability in the Humanities Grant and by three international travel grants from Cornell Mario Einaudi Center for International Studies.

Portions of this book's text were previously published in "Tools for an Efficient Witness: Deskilling Science and Devaluing Labor at an Agro-Environmental Research Institute," *HAU: Journal of Ethnographic Theory* 11 (2): 537–50; "Impact, Theory of Change, and the Horizons of Scientific Practice," *Social Studies of Science* 51 (1): 100–120; and "Amateurism and Our Common Concern for Biodiversity," *Made in China* 2 (2): 34–37.

Most of all, I am indebted to my family. To my parents and grandparents, without whose love and support I would never have been in the position to embark on this research. And to Nat and Alfie, for supporting me as I persevered with academia and this book despite the many reasons not to and for reminding me of the many things in life that matter so much more.

Acronyms

ANT	actor-network theory
CGL	Changing Global Landscapes project
CLEAR	Civic Laboratory for Environmental Action Research
CSIRO	Commonwealth Scientific and Industrial Research Organisation
IFF	Institute for Farms and Forests
IGO	intergovernmental organization
M&E	monitoring and evaluation
MELA	Monitoring and Evaluation for Learning and Evaluation
MOST	Ministry of Science and Technology
NGO	nongovernmental organization
NSF	National Science Foundation
NSFC	Natural Science Foundation of China
PRA	participatory rural appraisal
RCT	randomized-control trial
REF	Research Excellence Framework
RRA	rapid rural appraisal
SAPS	Songlin Academy of Plant Sciences
STS	science and technology studies
TAK	Toolkit for Agroecological Knowledge

SCIENCE INTERRUPTED

ETHNOGRAPHY WITH SCIENCE, DEVELOPMENT, AND BUREAUCRACY

"I'm really excited about this new project," Matt told me as we walked along a busy road. "Over the past five years, we've done a lot of talking about agroforestry, but this is a chance to really put things into action. In Myanmar, it could really make a huge difference too. The forests there have been devasted by logging, and for a lot of people, those forests are their livelihoods." Matt had spent the past five years working as a soil ecologist at the Institute for Farms and Forests (IFF) in southwest China. He and his colleagues had produced an abundance of scientific research establishing potential social and ecological benefits of biologically diverse agroforestry systems. For Matt, however, there was an important step that he and his colleagues needed to start making: one from understanding the potential benefits of agroforestry to establishing actual agroforestry systems.

Matt hoped that his new project, Agroforestry for Myanmar, would be an opportunity to make this step. With a multimillion-dollar grant from an international development donor, Matt's team would establish a series of experimental agroforestry plots in northern Myanmar. These plots would contain relatively simple agroforestry systems with trees spaced in grids or alleys and agricultural crops planted in between. The plots would begin life as sites for experimental research, but once research was underway, the same plots would also serve as demonstration farms. With training programs at these plots, Matt and his colleagues would try to convince land-users, policymakers, and nongovernmental organizations (NGOs) to adopt and support agroforestry across upland Myanmar.

A year or so later, on a somewhat quieter road near IFF's leafy suburban campus, Matt's hope had all but evaporated. "I remember when I used to get to do research," he told me forlornly. By this point, Agroforestry for Myanmar was mired in endless bureaucratic and institutional headaches: the project donor had retrospectively imposed an inflexible new evaluation framework that was incongruous with IFF colleagues' expectations for the project, local NGO partners were losing patience after months waiting for IFF headquarters to approve the transfer of project funds, and Matt was busy trying to persuade a key academic partner that it was neither possible nor necessary for IFF to purchase him a new four-wheel-drive vehicle. Amid this never-ending stream of headaches, the agroforestry systems Matt and his colleagues were so excited to plant, to research, and to promote receded into the background.

The tension between these two moments—a moment of hope and anticipation that gives way to one of frustration and dismay—is familiar to many of Matt's colleagues at IFF, as well perhaps to many academic colleagues elsewhere in the world. We are, on the one hand, motivated more than ever to put our research skills to work changing worlds beyond the university while, on the other hand, the institutions that employ and fund us operate in ways that leave ever-diminishing time for us to pursue that aspiration. These parallel tendencies are connected. Many of the proliferating bureaucratic evaluation regimes—the so-called audit cultures—that we bemoan in the sciences and humanities are themselves the result of efforts to promote and foster "impactful" research. Indeed, the mantra of maximizing "impact" was at the heart of the new donor monitoring and evaluation framework with which Matt and his colleagues were wrestling. The irony of this contradiction is captured by the commonplace refrain that audit regimes are corrosive of the very things that they are supposed to promote (Kipnis 2008; Power 1997; Shore and Wright 2015; Stein 2018). Certain aspects of the ethnography I present in this book underline this contradiction. As much as examining how bureaucracy and audit undermine the sciences, however, my goal is to explore the positive lessons scientific practices might draw from the logics of bureaucracies and from IFF staff's frustrations with them. I want to show how thinking with instead of only against bureaucracy might help us confront the task of rebuilding scientific practices for a more just and more livable world.

Thinking with Bureaucracy

By thinking with bureaucracy, I mean the application of a classic anthropological method that crafts analytic tactics out of ethnographic encounters. In this mode of anthropology, ethnography is not merely something to which we apply theories and concepts; it is also a practice through which we develop them. Some version of this approach to ethnographic theory is shared by anthropologists concerned with all manner of ethnographic objects and intellectual projects. It has given anthropology concepts from "obviation" (Wagner 1986) and "arbitrage" (Miyazaki 2013) to "decolonizing extinction" (Parreñas 2018) and "pluriverse politics" (de la Cadena 2015). In *How Forests Think*, to take one example, Eduardo Kohn (2013) challenges dominant Euro-American assumptions about what distinguishes humans from other beings by developing an innovative theory of nonhuman thought. For my purposes, Kohn's theoretical argument is less important than the methods that inspire him to reach it. In this respect, Kohn's concept of thinking forests emerges fundamentally from encounters with his ethnographic interlocutors—the Runa people of Ecuador's Upper Amazon. According to Kohn (2013, 94), his claims are "not exactly . . . ethnographic . . . in the sense that [they are] not circumscribed by an ethnographic context." But, Kohn continues, his claims "gro[w] out of [his] attention to Runa relations with nonhuman beings as these reveal themselves ethnographically." I would argue that this makes his theoretical claims *exactly* ethnographic. Ethnography in the sense I am interested in is not about circumscribing one's claims to a narrow context. It is about building analytic tactics out of encounters with the knowledges, practices, and experiences of our ethnographic interlocutors.

For many anthropologists and in many anthropological subfields, the idea of anthropology as a practice of thinking with our ethnographic interlocutors would go without saying. When it comes to certain objects of inquiry, however, anthropologists have been less inclined to adopt this method. This is perhaps especially so with the often-interconnected themes of audit cultures, bureaucracy, and international development. Rather than thinking ethnographically with bureaucratic and development practices, anthropologists have more frequently taken their theoretical inspiration from Foucault or actor-network theory (ANT) (for example, Escobar 1995; Ferguson 1994; Hull 2012; Merry 2016; Mosse 2005; Pels 2000; Rottenburg 2009; Shore and Wright 1999; cf. Yarrow and Venkatesan 2012). To be clear, my goal in departing from this approach is not to defend bureaucracy against unfair attack. Indeed, thinking with bureaucracy does not preclude simultaneously thinking against it. Nor, as importantly, is it to question the reality or significance of the violence, precarity, inequality, and injustice that Foucauldian and ANT-inspired approaches have allowed anthropologists to illuminate.[1] In emphasizing a more traditional approach to ethnographic theory, I want merely to suggest that for all that is wrong with bureaucracies—not least the ones that govern our own professional lives—there may nevertheless be creative potential in thinking ethnographically with them.

In this book, I explore this potential in relation to a planning and evaluation tool called Theory of Change. In basic terms, Theory of Change is a technique for developing a *theory* of how to bring about desired *change* in the world. It is often integrated with metrics for measuring the extent to which that change is or is not being achieved as anticipated. Theory of Change originated in development NGOs, but in recent years, it has become prevalent in diverse fields from philanthropy and international aid to universities and academic funding organizations. Australian academics may, for example, be familiar with it as a core component of the "impact framework" used by the federal research agency the Commonwealth Scientific and Industrial Research Organisation (CSIRO) (CSIRO n.d.). It is likewise a required component of certain funding opportunities offered by the US National Science Foundation (NSF) (NSF 2015, 2022).

Depending on who one asks at IFF (as well as which iteration of it you are asking them about), Theory of Change is either a bureaucratic headache keeping scientists like Matt from their research or a tool that can empower those same scientists to achieve the real-world impacts they aspire to make. Theory of Change can, moreover, materialize in a multiplicity of forms. In some instances, it is a colorful hodgepodge of notecards pinned to a board. In others, it is transposed into a neat spreadsheet replete with measurable targets to be met on a strict schedule. What is remarkable to me about Theory of Change, however, is the provocative framework that it provides for reflecting on three dimensions of scientific practice: the horizons that motivate research, the moments at which the future effects of research are anticipated, and how scientists imagine their agency playing out across time. It is, in anthropological terms, a framework for understanding and reimagining the temporalization of action and agency.[2]

Taking up Theory of Change's provocation, these dimensions serve as a framework for this book's analysis and argument. They are, for instance, the basis of my comparison between an IFF communications consultant named Alistair's plan to build a brand for agroforestry and a junior scientist named Jiaolong's more modest call for us to communicate our research to wider publics. In the former case, Alistair conceives a Machiavellian strategy for nurturing the eternal expansion of IFF's power and influence. In the latter, Jiaolong imagines an impact pathway that, like Alistair's, would entail scientists doing much better at interesting nonacademic audiences in their research. Unlike Alistair, however, her impact pathway anticipates a horizon at which scientists cede control of their work—a moment that anticipates a further horizon where research findings might be put to use in ways scientists neither preconceive nor control.

In Alistair's case, he casts his brand-building strategy as a "Theory of Change." Jiaolong makes no such connection. Since the start of her graduate studies, Jiaolong has been concerned with how scientific research might help people generate sustainable incomes from the rubber plantations that she had witnessed spread rapidly across her local landscape. "When I started my PhD," she told me, "I thought I was doing something that would help rubber farmers like my brother." This is an aspiration that predates her encounter with Theory of Change, and it is one that she would not herself frame in its terms. I nevertheless approach Jiaolong's proposal as an implicit Theory of Change. I do this not to suggest that Jiaolong's proposal is the same as Alistair's branding strategy or other explicit versions of Theory of Change. Indeed, her proposal is in important respects incompatible with many institutionalized versions of Theory of Change. Rather, framing Jiaolong's—as well as many of her colleagues'—ideas and practices as an implicit Theory of Change is one way in which I exploit the logic of Theory of Change as my own analytic framework—a framework that can, among other things, help us understand latent alternatives to institutionalized imaginations of impact.

Thinking ethnographically with Theory of Change is in this respect distinct from making a case for its embrace within the academy. My argument is that bureaucratic forms like Theory of Change might provide productive problematizations—for example, "What is the horizon that motivates research?"—not that they necessarily offer ideal solutions. In certain respects, thinking with these problematizations can help us articulate what is wrong with Theory of Change as a way of doing science. We can, for instance, follow Theory of Change's lead by comparing evaluation regime–imposed horizons with the horizons that implicitly motivate IFF scientists. This comparison can help illuminate the limitations of certain conventional research practices, but it can also help us question the very peculiar expectations that Theory of Change–based evaluation regimes generate. In adopting this ambivalent orientation toward bureaucracy, I take inspiration from feminist and decolonial science and technology studies (STS) scholars who have in a variety of ways sought to repurpose ostensibly objectionable scientific concepts and practices to the service of critically analyzing and transforming the sciences (for example, Fanon 1963; Grosz 1999, 2005; Haraway 1991; Harding 1991; Murphy 2021; Roy 2008, 2018; Subramaniam and Willey 2017b).[3] Thinking in terms of bureaucracy-inspired problematizations may at times give us pause to ask whether bureaucracy (or at least some transformed version of it) might have a positive role to play in remaking the sciences. At other times, these problematizations will help us conceptualize what is wrong with the audit cultures that govern contemporary science, academia, and international development. Most importantly, thinking with bureaucracy can help us reflect on the aspiration that Matt and Jiaolong share with colleagues at IFF and elsewhere: that of doing science in ways that might help transform the deeply troubled worlds we inhabit.

Bureaucracy as Interruption

Many of the bureaucratic and scientific practices I describe in this book will resonate with experiences beyond one office in southwest China. Indeed, a key
goal of this book is to speak to challenges, frustrations, and aspirations that are
relatable across disciplinary and national boundaries. Nevertheless, this book,
its concepts, and its argument emerge fundamentally out of ethnographic fieldwork at a specific place: the Institute for Farms and Forests' China office. This is
not an ethnography of bureaucracy and Theory of Change in the abstract but
one of and with the perspectives of IFF colleagues as they encounter them in
their professional lives.

IFF is a global agroforestry research for development organization with its
headquarters in East Africa. IFF established an office in Songlin—a city in southwest China—almost twenty years ago. Originally a country office responsible
for activities within China, by the time of my fieldwork (2014–16), it had ex-
panded to take on a regional role with active projects elsewhere in East, Central, Southeast, and South Asia. Broadly speaking, IFF's mission is to produce
scientific knowledge that will inform and enhance equitable and sustainable rural development—it pursues "research for development." But the question of
what IFF does and why it does it is a large part of what is at stake in the divergent ideas of science and impact that I explore in this book. IFF's projects at the
time of my field research included Eco-Friendly Rubber, a project to develop sustainable and environmentally friendly rubber cultivation practices for southwest China and neighboring Southeast Asia. This project operated in tandem
with the Lengshan Multi-Stakeholder Platform, an initiative set up in Lengshan
Prefecture, southwest China, as part of a broader global project to facilitate mul-
tisector collaboration in tropical agriculture. The Qingshan Agroforestry project held similar aspirations to Matt's Agroforestry for Myanmar project, only
in this instance focused on Qingshan, a small township in southwest China. A
broader Asian Fungi project involved numerous subprojects from taxonomic
macrofungal surveys to social science research on mushroom harvesting.

These projects were led and implemented by an office of around twenty research staff, ten corporate services staff, and forty or so affiliated graduate students. Approximately half of the staff and students are Chinese, with large
contingents from elsewhere in Asia as well as from Africa, Europe, and North
America. This staff is diverse in terms of their expertise and training as well as
in the aspirations and goals they hold for themselves, for IFF, and for environmental research more broadly. This is evident in, among other things, their diverse responses to new institutional mandates to employ Theory of Change as a
method for planning and evaluating IFF's work.

Like many at IFF, my first encounter with Theory of Change came in the form of an "outcomes thinking" workshop run by a headquarters-based senior scientist named Lesley. I remember sitting through the first morning of her workshop wondering whether what Lesley was telling us about proactively communicating research findings (as opposed to simply sticking them in an academic journal or policy brief and leaving it at that) was not already obvious to everyone in the room. "Isn't what Lesley is saying more or less what IFF already does?" I asked Matt at lunch. "What is she saying that's new?" "This is really important," Matt responded. "What she is describing is a long way from what we've been doing, and we really need to start doing what she is teaching us." Part of my misapprehension, I realize in retrospect, was my incorrect assumption of a natural affinity between the applied agroforestry science that many already practiced at IFF and outcomes- or impact-focused science. I did not yet appreciate the difference between a research practice that anticipates the utility or application of new knowledge in the hands of undefined future users (IFF colleagues' existing approach to applied science) and a research practice that plans for knowledge to be deployed to predefined ends by predefined users (outcomes thinking). For Matt, moreover, even applied science had been relatively peripheral to his work. In this respect, whereas Jiaolong had always imagined a close connection between her research and transformations in agroforestry practices, for Matt such a direct connection was more novel. Having spent much of his previous five years at IFF taking significant satisfaction (and institutional reward) from the publication of what he called "fundamental research," Matt appreciated Lesley's offering as an opportunity to start realizing the potential of his research to generate real-world change.

Bob, a more junior colleague—and the only Songlin-based staff member familiar with Theory of Change prior to Lesley's workshop—gave a slightly different assessment of what was new in Lesley's workshop. As we climbed the stairs to join colleagues for a morning coffee break, Bob explained to me that "everyone at IFF already has an implicit Theory of Change. We all already work according to a set of assumptions about how one gets from research to change in the world." A key value to Lesley's workshop, Bob continued, was that it forced his colleagues to reflect more explicitly on those assumptions. Theory of Change, Bob suggested, is a reflexive process that can help IFF colleagues rethink and revise how they go about their work. This book's attempt to think with Theory of Change follows through on Bob's basic idea here—that Theory of Change can help us make explicit and reflect on existing imaginations of the processes that connect research to the worlds beyond the academy.

Bob and Matt's positive attitude toward Lesley's workshop was not shared by everyone. Ruyue, a postdoc in Matt's soil biology group, was one of several colleagues who attended only the first morning of the two-day event. She later told me, "I've heard it all before," before joking that I would "get used to this sort of workshop soon enough." What Ruyue had heard before was not Theory of Change specifically but headquarters scientists flying in for a couple of days to share ideas that held minimal relevance to IFF-China's work. A year—and several encounters with Theory of Change—later, Matt had adopted a similar view. When word got around that Lesley was planning a follow-up workshop, Matt joked with colleagues about what field research they could reschedule to make sure they were out of the office for the workshop.

There are important differences in how Matt and Ruyue experience Theory of Change and other externally imposed mandates. These colleagues share, nevertheless, a concept for thinking about such things: for most IFF staff, Theory of Change is just one of the many *mafan*, or "headaches," that are part and parcel of professional life. A dictionary translation of the Chinese *mafan* would be a "trouble" or a "nuisance." In the office's common English parlance, the term is often left untranslated, operating as a loanword even among colleagues like Matt who speak very little Chinese. When it is translated, *mafan* usually becomes "headache," "pain in the backside," or some more colorful equivalent.[4] This is a label IFF colleagues apply in a range of bureaucratic contexts, flattening distinctions between what we might otherwise imagine as quite distinct challenges: drafting a Theory of Change, applying for a visa, completing donor project reports, procuring research permissions, filing expense reports, satisfying Chinese anticorruption procedures, and so on.

This notion of headache contains an implicit Theory of Change of its own: a particular way of imagining the temporality of bureaucratic and scientific action and agency. Rendered as a headache, IFF colleagues experience bureaucratic work (itself a fluid category) as a momentary *interruption* to meaningful work. Confronted as a headache, the interrupting moment is animated exclusively by the question of how the nuisance in hand can be overcome as quickly and efficiently as possible.[5] As IFF colleagues understand it, moreover, the in-the-meantime labor of overcoming a headache is something that will have little or no effect on the scientific work that will resume once the headache is out of the way. Headaches do not restrict or transform scientific practice; they merely put it on hold.

One thing that is striking about IFF staff's rendering of bureaucracy as a headache is that it suggests a very different kind of analysis to that offered in social scientific analyses of academic audit cultures or of monitoring and evaluation (M&E) regimes in international development. A central question in studies of university audit regimes has been "what audits and rankings bring into focus and what they render invisible or unsayable" (Shore and Wright 2015, 422). Similarly, in the field of international development, David Mosse (2005) analyzes development projects as a "system of representations," and Richard Rottenburg (2009) describes development project evaluations as a production of "immutable mobiles" for subsequent recombination into representations that serve the hegemonic interests of Western donors. Such focus entails social scientists drawing insight out of the incongruity between what audit and bureaucracy make visible or legible on the one hand and the alternative representations that ethnographers or their research subjects might make of the situation on the other.[6] For IFF staff, by contrast, the paucity or opacity of the representations that bureaucratic procedures produce is of little interest. In the context of university audit cultures, understanding bureaucracy as a means of making academic work visible is suggestive of a mode of critique that highlights the failure of audit to provide an adequate medium of representation (Giri 2000). This in turn can imply that our challenge should be to craft regimes that can make academic performance visible on academics' own terms (Shore and Wright 2015). But what kind of critique and what kind of imperatives for transformation are implied by an understanding of bureaucracy as a mode of interruption?

In a straightforward sense, the concept of a headache might suggest that what is wrong with bureaucracy is that it interrupts at all. Indeed, the notion of headache indexes an ideal of scientific work shorn of nuisance distractions such as donor accountability and institutional regulation. Rather than indulge the fantasy of scientific practice free from interruption, however, I consider how central components of scientific practices such as peer review, PhD defenses, and participatory workshops are, no less than bureaucracy, forms of interruption. Unlike headaches, however, these are interruptions in which IFF staff anticipate opportunities for feedback that can positively transform the future direction of a project. Exploring the various ways in which IFF scientists actively seek opportunities for their work to be interrupted by communities within and beyond the sciences, I argue that interruptions are vital to scientific practice—both as it exists now and in the more collaborative and democratic forms of science that we might wish to build. These forms of interruption are moments of willful abeyance of agency (Miyazaki 2004), ones that anticipate our interlocutors shaping our research in unpredictable ways. The critique of academic bureaucracies from this perspective would be that the interruptions these bureaucracies generate fail to provide opportunities for generative abeyance. And if the critique of bureaucracy in the academy were to focus on the inadequacy of the interruptions that bureaucracies generate, our challenge might be to imagine not alternative forms of visibility but alternative modes of interruption.

Trust and Vulnerability

At the time of my research, IFF was officially registered with the Chinese state authorities as an international research organization. This is an arrangement common to the numerous international agricultural research organizations operating in China. More unusually, the office is also legally constituted as a research center within a prestigious public research institute named Songlin Academy of Plant Sciences (SAPS). The same office therefore exists both as an international organization and as a domestic public research institute. This dual identity allows some advantages but can also mean double the bureaucracy. As a colleague named Tao put it, having feet in two institutions means "getting screwed from both sides."

Tao grew up in Songlin before heading overseas to study and work in the fields of rural development and environmental sustainability. Returning home to take up a senior position within IFF's corporate services team, Tao was attracted to the organization by its mission to build sustainable socio-ecological futures. Like many of her colleagues, however, she was frustrated that so much of her time was devoted to bureaucratic headaches that did nothing but get in the way of that work. Though these headaches came from "both sides," Tao saw one important difference between Chinese bureaucracy and that of IFF's global headquarters. Dealing with the latter, she told me, "There is at least the possibility of flexibility." When she encountered IFF regulations that imposed an unreasonable or impossible burden on the China office's work, Tao knew she could speak to a colleague at headquarters and try to negotiate an exception. This, she told me over a lunchtime coffee, "could never happen with the Chinese side": even when following regulations to the letter would make vital scientific activities like field research and academic conferences impossible, Chinese administrators would force IFF to comply.

Echoing Matthew Hull's (2012) ethnography of Pakistani bureaucracy, Tao suggested to me that officials demand strict adherence to (or at least the appearance of adherence to) document-mediated procedures in order to diffuse the possibility of personal responsibility for the activities they authorize. Responsibility, she told me, is feared because of the vulnerability it generates: if a bureaucrat helps bend the rules, they risk later rebuke from their own overseers. Labeling this phenomenon a "refusal to trust," Tao equates refusing trust with evading vulnerability. Conversely, her analysis implies that to trust might mean to embrace the vulnerability that comes with entanglement in the affairs of another. This is a concept of trust that can help us appreciate instances in which Chinese officials seem to trust IFF colleagues in precisely this manner: when they willingly entangle themselves in IFF's activities and in doing so make themselves vulnerable to the uncertain outcomes collaboration might bring.

As well as drawing on Tao's notion of trust as vulnerability to appreciate IFF's collaboration with state officials, I argue for the willful embrace of vulnerability as a model of trust for scientific practice. In making this argument, I borrow from the ambivalent conceptions of vulnerability developed in recent feminist scholarship (Butler 2015; Butler, Gambetti, and Sabsay 2016; Parreñas 2018; Puig de la Bellacasa 2017). In this literature, "Vulner*ability* is not just a condition that limits us but one that can enable us. As potential, vulnerability is a condition of openness, openness to being affected and affecting in turn" (Gilson 2011, 310). This ambivalent perspective does not deny that we must continue to challenge the many institutionalized forms of vulnerability and precarity that professionals in the sciences experience at the hands of audit cultures—or, relatedly, because of culture wars, funding cuts, and labor casualization. It does, however, allow us to differentiate the work of managing our vulnerability from "valuing invulnerability as a context-free unequivocal good" and encourages us to grapple with the "repercussions of closing oneself to certain kinds of relations and situations" (Gilson 2011, 323). It might help us, moreover, to encounter and respond to the destructiveness of bureaucracy and audit cultures while keeping in focus the simultaneous necessity of sustaining and proliferating conditions of openness and relatedness that will be crucial to transforming scientific practice.

In a certain respect, this focus on vulnerability is a logical extension of meaningful interruptions as moments in which one cedes agency. When IFF colleagues invite feedback from scientific peers or rural communities, such invitations are meaningful only insofar as they embrace the vulnerability of their work to the uncertain responses of their interlocutors. Focusing on moments in which IFF colleagues render their work and their knowledge vulnerable, I demonstrate the vital role that vulnerability can play in scientific practice. Tying trust to vulnerability, moreover, suggests a particular orientation to a challenge that animates many in the contemporary academy: that of rebuilding public trust in science and expertise. Equating trust with vulnerability would require us to reject tired "trust us, we're experts" models of science-society relations and would allow us to differentiate building trust from claiming autonomy or epistemic authority. The argument, in short, is that if we want communities beyond the academy to trust us, we may need to ask: How can we make ourselves vulnerable to them?

In thinking about the significance of and possibilities for vulnerability in scientific practice, I draw on STS practitioners who have long disrupted the invulnerable authority of Science and pushed for more collaborative and inclusive research practices. These are scholars who have critiqued detached, aperspectival, and mechanical models of objectivity and who advocate instead for practices that embrace the entanglement of sciences in the diverse human and nonhuman worlds that we both inhabit and shape (Latour 2004a; Haraway 1991; Harding 2015; Liboiron 2021; Roy 2008). This literature informs this book's critique of value-for-money agendas that are tied to problematic understandings of scientific rigor and objectivity. As significantly, the way in which feminist STS has foregrounded and innovated alternative modes of scientific practice guides my ethnographic attention to occasions in which IFF colleagues operate outside and in conflict with the dominant logics of contemporary agricultural and environmental research for development. In chapter 7, my attention to Jiaolong's call for scientists and government officials to narrate their personal investments in Eco-Friendly Rubber is, for instance, inspired by feminists who celebrate and foreground storytelling and the personal in scientific practice (Keller 1985; Haraway 1991; Hubbard 1988; Roy 2018; Prescod-Weinstein 2022; Subramaniam and Willey 2017a). The subsequent rejection of Jiaolong's proposal is indicative of familiar logics and assumptions about Science that remain hegemonic in institutions like IFF. More important for my purposes, however, is the fact that even within such institutions we find colleagues like Jiaolong imagining and practicing sciences in ways that gesture toward more inclusive and collaborative alternatives. Jiaolong's turn to the personal is, in other words, one example of ways in which IFF scientists are already making research vulnerable despite the institutions they inhabit. One could doubtless find more sustained and more systematic examples of science vulnerable to communities beyond the academy—in, for example, the Civic Laboratory for Environmental Action Research's (CLEAR) pollution research or Deboleena Roy's molecular biology (Liboiron 2021; Roy 2018).[7] Nevertheless, starting from Sandra Harding's (1991, 302) premise that successor sciences are "already being developed" in a multitude of settings, I engage Jiaolong as one of many colleagues who might inspire more vulnerable research.

Collegiality and Participant Observation

During my field research, I worked for IFF as a research assistant. This allowed me to participate in several of the projects I describe. In the Agroforestry for Myanmar project, I shared responsibility with another junior member of staff for drafting donor reports, including the Theory of Change spreadsheet described in chapter 1. In the Qingshan Agroforestry project, I served as a Chinese-language interpreter and workshop facilitator for the two soil scientists leading the project. For the Asian Fungi project, my role was to complement IFF's ongoing biophysical research by designing and implementing a household survey on local mushroom-harvesting practices. In the Eco-Friendly Rubber and

Lengshan Multi-Stakeholder Platform projects, I participated in planning meetings and attended stakeholder meetings as a representative of IFF. I was a more fleeting participant in various other projects and workshops, often as a notetaker or as an English-language editor.

On one occasion, an American scientist attached to one of IFF's sister organizations asked with a good-humored chuckle: "Is working for the organization you are researching not a conflict of interest?" "Yes," I responded, "but not because of the money." The modest salary IFF paid me while I was conducting field research for this book is not any incentive to write a book-length advert for the organization's virtues or to obscure its many flaws.[8] Nevertheless, there are important ways in which my personal and professional interests are, if not conflicted, most certainly entangled with those of my colleagues at IFF. These colleagues are people with whom I shared not only an office but also dinners, beers, field trips, and weekends away. Many IFF staff became and still are friends as well as colleagues. Importantly, I share many IFF colleagues' aspirations for a world in which science helps us achieve healthier and more equitable rural landscapes. This does not mean that I intend for this book to validate my colleagues' work (as if they needed or wanted me to do so), but I do hope that it reflects the empathy and respect that I owe them. "Conflict of interest" is a bad way to describe debts of this kind, but as is the case for most ethnographers, I neither claim nor desire the detached objectivity with which scientific methods are stereotypically associated. Personal and professional entanglement in the worlds that we study is a vital component of anthropological research.[9]

The fact that I worked as an IFF research assistant throughout my fieldwork is one reason I refer to IFF staff as "colleagues." For the two years I spent conducting my fieldwork, they were colleagues in the most literal and immediate sense. In a certain respect, this kind of collegiality is a fact of any participant observation research. Most anthropologists' methods rely on the development of enduring relationships with ethnographic interlocutors. When it comes to ethnographies of scientific communities, there may also be a sense of collegiality that precedes fieldwork: we already share membership in transnational academic communities. My use of the term "colleague" references this second sense of collegiality too. It reflects the fact that even without having IFF in common, many of its staff are people who I might well encounter at academic conferences or seek to speak to through publications like this book. Of course, there is much that divides anthropology from the natural and social scientific disciplines in which IFF research staff are trained, not least the fact that as an anthropologist I enjoy a luxury not afforded to any of my colleagues: that of stepping back from fieldwork/employment at IFF to write this book. But, as I allude to in the chapters that follow, there are many ways in which anthropology is not quite as different from agricultural and environmental sciences as stereotypes of the natural sciences might have us believe.

Something that sits awkwardly with my commitment to academic collegiality is my use of pseudonyms for personal as well as institutional and place names. This anonymity is important in part because I describe bureaucratic workarounds that are not always in the spirit or letter of the operative institutional rules. Notwithstanding the use of pseudonyms, it remains possible that readers familiar with environmental research in China will recognize some of the parties involved in my research. To reduce the recognizability of individual staff members, I am deliberately vague about their ethnic and national identities as well as about certain aspects of their careers. To minimize the risk of revealing specific, identifiable instances of the bureaucratic workarounds described in this book, I have furthermore fictionalized aspects of my ethnographic account, and I have limited my ethnography to examples of relatively commonplace bureaucratic tactics to which administrators routinely turn a blind eye. All of the preceding steps aim to decrease the possibility that anyone in a position of bureaucratic authority would have either the motivation or the ability to decipher any specific administrative infractions described in this book. The flip side to these steps, however, is to make it impossible for me to cite my IFF colleagues as academic colleagues. In chapter 1, for instance, I quote from a blog in which Jiaolong articulates an argument for scientists to share their ideas and findings with communities beyond the academy. That blog has been just as important an inspiration for this book as any anthropology or STS text, but I do not provide citations to Jiaolong's or her colleagues' publications because doing so would undermine my efforts to anonymize IFF and its staff.

Though I have not fulfilled the usual obligation to cite, I have attempted to follow through on other aspects of academic collegiality. This has included sharing drafts of this and other writing with some of the colleagues I worked alongside in China. For a member of the corporate services team at IFF who is less familiar with academic English, I also attempted to summarize aspects of my analysis in Chinese.[10] IFF colleagues' responses to my ethnographic writing have been at times critical but always generous. I benefited greatly, for instance, from a colleague named Lesley's commentary on misapprehensions of Theory of Change that I unwittingly shared with other IFF-China staff. This was an observation Lesley made reading a draft of an earlier article, aspects of which are reproduced in chapter 1 (McLellan 2021a).

Another IFF colleague's response to that same draft has been especially informative in my thinking about this project. Via email, Jacob explained to me that my article "spells out nicely what I have felt for a long time but in a less articulated way about the discrepancy of the science way and the development way." In many respects, this was (and I know was intended to be) a complimentary thing to say about my work. At the same time, however, this comment led me to question my aspirations for this ethnographic project—my own implicit Theory of Change. Jacob's comment suggested that what my ethnography had achieved was an articulation of bureaucratic frustrations that he already knew too well. As I had described in the draft article he was responding to, Jacob is no fan of the "development way" epitomized by Theory of Change and was (for good reason) exhausted by the obstacles that funders were placing in the way of projects like Agroforestry for Myanmar. Illuminating what is wrong with this development way is important. Indeed, it is something that I hope this book achieves. Nevertheless, Jacob's email left me with the sense that all I had managed was yet another account of the evils of what we might also call the "audit culture way" or the "impact way." Jacob's comment reminded me that my goal for this project had been to do more than show that bureaucracy can get in science's way; it had been to reflect on scientific practice itself.

Pondering Lesley's and Jacob's emails in relation to this ethnographic project, I began to see the obligation to challenge—or, more precisely, to *interrupt*—one another's thinking as a key part of scientific collegiality. In this respect, Lesley's feedback had interrupted my thinking about Theory of Change by illuminating a dimension to its logic and implementation that I had previously missed. Just as valuable, Jacob interrupted my thinking by highlighting the limited ambition of my writing. Returning Jacob's collegiality, I realized, might mean writing this book in a way that might interrupt his thinking in as productive a way as he and his colleagues have interrupted mine. This is the spirit of collegiality in which I have tried to write this book, one in which empathy and generosity are crucial but equally in which I hope to interrupt how colleagues at IFF and elsewhere approach their bureaucratic and scientific work.

Chapter Overview

In chapter 1, I introduce three iterations of Theory of Change that I encountered at IFF. Drawing on the frustrations of IFF staff coming to terms with Theory of Change, I show how institutional demands for "impacts" impose temporalities that fundamentally shape the kind of actions and relationships it is possible to imagine and to practice. In addition to exploring the strictures that Theory of Change imposes on scientific practice, I set up Theory of Change as something for us to think with—as a framework that can help us understand scientific and bureaucratic practices at IFF as well as reflect on the wider task of remaking scientific practice.

In chapter 2, I describe how IFF staff routinely understand bureaucratic work from procuring government research permissions to submitting grant reports as a "headache." Whereas advocates of Theory of Change promote it as a tool for generating momentum toward impactful research, rendered as a headache bureaucratic practices like Theory of Change are reduced to nothing but a meaningless interruption. In addition to elaborating this negative sense of an interruption as a waste of time, I juxtapose bureaucratic headaches with IFF research practices during which interruptions of various kinds are valued, productive, and indispensable parts of the scientific process. Drawing attention to the quality of interruptions that bureaucratic and scientific practices make to research projects, I shift the problem of audit and academic bureaucracy from one of visibility and representation to a question of the quality of interruptions that bureaucracies generate. I also argue that thinking in terms of interruption will allow us to disentangle the problem of trust in the sciences from the goal of scientific autonomy and to reconsider what we mean by the "trust" that we seek as scientists and academics.

Chapter 3 shifts attention toward IFF colleagues' navigation of Chinese state bureaucracies, including documentary workarounds to headache-inducing anticorruption measures, and the creative labor of local staff to secure access to research sites for their foreign colleagues. Taking my lead from Tao's description of the obstructiveness and inflexibility of Chinese bureaucrats as a "refusal to trust," I outline a concept of trust that is tied to a willingness to render oneself vulnerable, and I elaborate how desires to evade vulnerability give rise to bureaucrats' insistence on rigid adherence to formal documentary procedures.

Having explored IFF colleagues' critiques of bureaucratic vulnerability evasion in chapter 3, in chapter 4 I focus on instances where IFF staff display sympathy toward the vulnerabilities that state officials encounter. This includes an instance where Jiaolong found a government official opening himself to vulnerabilities that he might otherwise have evaded. Borrowing from recent feminist and anthropological considerations of the productive possibilities of vulnerability (Butler 2015; Butler, Gambetti, and Sabsay 2016; Gilson 2011; Parreñas 2018), I suggest that this official's openness to vulnerability lays a pathway for collaboration and trust.

In chapter 5, I examine the value-for-money logic that undergirds research for development tools. "Tools" constitute a diverse genre, but they often share several core characteristics: a drive for quantifiable certainty, the deskilling of data collection and project implementation, and a pretense to scale and speed. These are characteristics that I contrast against the generative anxiety of a project—Qingshang Agroforestry—that is at odds with certain of the hegemonic logics that tools exemplify.

Continuing with the theme of research for development tools, chapter 6 explores some of the embodied skills and tacit knowledges that deskilled tools pretend to do away with—from the feel for the environment developed by a mycologist with an enviable eye for hidden mushrooms to the tacit knowledge of a survey enumerator who is a master of collecting household survey data in rural China. In doing so, I illuminate the "playful modes of inquiry" (Dumit 2021) that are as crucial to the natural sciences and quantitative social sciences as they are to anthropology but that are suffocated by efforts to provide value for money through rapid and scalable research. As well as highlighting the consequences of value-for-money logics for research staff, I argue that tools exacerbate the routine invisibility of "nonacademic" staff whose intellectual contributions to the sciences are as invaluable as they are underappreciated.

In chapter 7, I further examine aspects of IFF's work that fall outside the suffocating value-for-money imperatives discussed in the preceding chapters. This includes the Qingshan Agroforestry project's freedom from the kinds of audit processes that govern most of IFF's other projects. This is a project that I introduce in chapter 5 as an example of a "scen[e] that exceed[s] or escape[s] 'professionalization'" (Tsing 2015, 285): as a project that casts aside the pretenses to value for money and to hubristic certainty that characterize dominant models of impactful research. I argue that one effect of this escape was for the project team, anxious about the uncertainty and incompleteness of their knowledge, to make their project plans repeatedly and willfully vulnerable to the objections and interruptions of local farmers. At the same time, however, I suggest that stepping outside our usual professional constraints entailed evading other kinds of vulnerability—not least accountability to those who fund our work and to the professional peers who might elsewhere serve as our "community of critics" (Strathern 2006). In this respect, chapter 7 is concerned with the tension between parallel impulses to generate vulnerability to local farming communities and to evade vulnerability to publics and peers beyond Qingshan.

In the conclusion, I place this book's Theory of Change into relation with anxieties that universities and the sciences are under threat from, among other things, anti-expert populism, neoliberalism, and audit cultures. In response to the destructive precarity of the contemporary university, I make the case for building scientific practices that cultivate vulnerability and that proliferate opportunities for research to be interrupted by broader publics.

THEORY OF CHANGE

Though I had arranged to take up a twelve-month internship at IFF many months before my arrival in Songlin, my would-be supervisors had always insisted that precisely what I would contribute as an intern would be "something to work out once you arrive." My arrival on campus in 2014 met with a further deferral: the office's senior staff were all away at IFF's annual global meeting in East Africa. The absence of senior staff made for a more laid-back environment than the somewhat tense atmosphere that the presence of the institute's director, Professor Yin, often seemed to create. My timing was therefore fortuitous in the opportunity it provided to catch up with colleagues I had met during an earlier visit in 2012 as well as to get to know some of the new faces. As colleagues in China enjoyed a week of relative respite, however, tense conversations were playing out on the other side of the world about the future of IFF as a global research for development organization and about the China office's contribution to the organization's evolving strategy.

On his return from East Africa, a senior scientist named Matt asked me if I had time to meet. We sat down in the top-floor conference room, and Matt began to explain the trouble IFF-China was in with global headquarters. Yin, Matt explained, had arrived at the annual meeting confident of a warm reception. Despite the China office being one of IFF's smaller regional offices, the volume and quality of papers that Yin and his colleagues had published in the preceding year exceeded the achievements of most of the other IFF offices dotted across Asia, Africa, and Latin America. But far from the congratulations Yin had anticipated, Matt told me that IFF's global leadership rebuked Yin for neglecting "outcomes."

Matt was somewhat vague in his understanding of exactly what IFF's new "outcomes-orientated" emphasis entailed, but the annual global meeting had convinced him that it was something he would have to start taking seriously. It was also an approach that Matt was certain would require some social science, a concern for Matt as his soil biology research group was made up exclusively of natural scientists. He explained to me that he was hoping to recruit social scientists to his team and asked me if I would be one of them. In the new "research assistant" role that he imagined for me, I would spend my time in rural southwest China running community workshops and questionnaires, doing things like asking communities how they understand a particular farming practice, figuring out what we can learn from indigenous knowledge, and finding ways to get farmers and foresters across the region to adopt best practices. Matt explained—to my relief—that this was not something he imagined as a solo effort. Much of what he had in mind would be in collaboration with more experienced social scientists at IFF headquarters and elsewhere, but he wanted someone in his team who could both help him make sense of what his new social science collaborators were after and carry out some of the local fieldwork these collaborations would require.

Less than a week into my fieldwork, this was a discomforting but also exciting position to find myself in. On the one hand, it had always been my intention to put myself at IFF colleagues' disposal. Matt's job offer was in certain respects exactly what I had been after: an invitation to be a *participant* observer. On the other hand, however, Matt's broad and vague expectations of "social science" jarred somewhat with the skill set I imagined myself to hold as an aspiring anthropologist. Prior to my meeting with Matt, I had anticipated intern responsibilities like language editing and helping with literature reviews. In such tasks (which I did also carry out), my primary value was as a native English speaker surrounded by colleagues compelled to publish research in what was their second or third language. Accepting Matt's job offer meant committing myself to a much less familiar form of work: joining Matt and his team in learning and practicing an entirely new mode of expertise. When I explained my unfamiliarity with whatever outcomes-focused work might be, Matt chuckled. "Don't worry," he told me. "It'll be a steep learning curve for us all."

One of the first steps on this learning curve was a workshop on "outcomes thinking" delivered by a visiting participation and impact scientist from IFF's international headquarters. During her workshop, Lesley introduced a planning and evaluation framework called Theory of Change. Most IFF colleagues had never heard of Theory of Change before Lesley's visit in late 2014, but by the time I departed IFF two years later, it was a ubiquitous presence. As I describe in this chapter, it is a framework IFF colleagues subsequently encountered in an international donor's monitoring and evaluation (M&E) regime and in a talk on brand building delivered by an IFF communications consultant. Borrowing the peculiar sensibilities of Theory of Change as my own comparative framework, I explore two further models of scientific practice at IFF: research as it is conventionally imagined by IFF colleagues and a plea made by one IFF colleague for scientists to proactively interest broader publics in our research.

Comparing these five Theories of Change and implicit Theories of Change, I develop three points: First, I highlight that the normative stakes in new institutional agendas for impact cannot be reduced to a straightforward shift toward research that is more oriented to practical, real-world effects. In this regard, the science done by IFF colleagues had long been oriented toward practical, real-world problems like conservation, climate change mitigation, and poverty reduction. What impact agendas transform is—in the context of agricultural and environmental sciences at least—more subtle. They transform the structure according to which effects in the world are anticipated and the process by which scientists are to manage the pathway from research to impact. Three aspects are especially significant to understanding the various orientations to impactful research that I encountered at IFF: the horizons that motivate research, the moments at which future effects are anticipated, and how scientists imagine their agency playing out across time. Second, by exploring the diverse ways that Theory of Change is applied—even within a single organization—I draw attention to the multiplicity of forms that institutional agendas for promoting impactful research can take. Third, taking my lead from the peculiar sensibilities of Theory of Change, I explicate what I refer to as IFF scientists' "implicit Theories of Change"—their alternative imaginations of how contemporary science does and should generate effects in the world. In doing so, this chapter sets up Theory of Change as a theoretical framework that I borrow for my own analysis of scientific practice.

Theory of Change 1: The Aspirational Future of Outcomes Thinking

During her workshop, Lesley introduced a version of Theory of Change that makes an *aspirational future* the new driving force for IFF's collective work and that demands concrete, albeit provisional, plans for who will take up IFF's research and to what use. Central to Lesley's outcomes thinking workshop was the introduction of a new set of categories with which she asked scientists to conceptualize an impact pathway and Theory of Change flowchart for their research. Lesley put significant energy into explaining the differences between "outputs," "outcomes," and "impacts" as well as between "next-users" and "end-users." In this scheme, an output refers to a deliverable such as a policy paper or a workshop. According to Lesley, agricultural researchers have too often targeted outputs—especially in the form of scientific publications and policy papers—and then thought no further. Lesley, however, encouraged her colleagues in China to imagine pathways for their work that extend beyond the horizon of knowledge production and ultimately onto impacts for specified end-user groups. This involves IFF activities targeting changes in the "knowledge, attitudes, and skills" of next-users such as policymakers, donors, or NGOs.[1] These changes would in turn give rise to outcomes such as changes in policy or the adoption of new development projects that would benefit some ultimate beneficiaries—the end-users. Thus, an impact pathway might map a causal chain from IFF research on rubber cultivation, to a specified change in policy (an outcome) by a local government (a next-user), to an increase in crop yields (an impact) for rubber farmers (an end-user).

One aspect of outcomes thinking that Lesley emphasized is that scientists cannot undertake the process of generating outcomes after the fact of scientific knowledge creation. According to outcomes thinking, shaping the practices of next-users is most effective when scientists engage next-users from the beginning of a project. This implies a collaborative agenda and is closely connected to broader trends for multistakeholder meetings and platforms (Brown and Green 2017; Hall and Sanders 2015; Rabeharisoa and Callon 2004; Sanyang et al. 2016). Multistakeholder platforms bring together people from diverse backgrounds to work to address a common set of problems. Along these lines, IFF was already an initiating partner in the establishment of the Lengshan Multi-Stakeholder Platform—a collaborative project that aimed to address social and ecological challenges in the tropical Chinese prefecture Lengshan. Rather than decide on their research focus autonomously, platform facilitators asked IFF scientists to attend platform meetings along with stakeholders from business, government, and other research organizations—all of whom are potential next-users. From an outcome thinking perspective, this platform offered an opportunity for scientists to identify key next-users and to shape their research approach to the demands and needs of these next-users. By continuing this collaborative engagement throughout the research, scientists would, so the platform organizers told them, enhance their ability to mobilize the interests of next-users and thereby their prospects for effecting desirable changes in next-users' knowledge, attitudes, and skills.[2]

Having explained this new outcomes-focused philosophy and vocabulary, Lesley introduced a tool for planning a specific pathway for impactful research: Theory of Change. Often organized as a flowchart, activities are placed at the bottom. Arrows connect these activities to outputs and then to the outcomes that the outputs will generate. At the very top of the Theory of Change flowchart is a "vision." This vision is an image of the world that one wishes to bring into being—the aspirational future one imagines as the cumulative effect of the impacts one will generate. Building a Theory of Change, one begins at the top—with the vision—and then works backward toward the activities.

Working with giant colored cards, pinboards, and easel paper, Lesley invited her colleagues to draft a Theory of Change for IFF-China. We began with a vision statement: "Rural landscapes sustain healthy and culturally diverse ecosystems that ensure food security and provide health, wealth, education, cohesion, and equity. Public, private, and non-profit institutions support policies, investments and interactions that integrate sustainable land management with healthy urban-rural food and ecosystem linkages."

The utopianism of this vision did not escape IFF staff. One workshop participant joked that once we have reached this point, we could all go home. Though utopianism appears inappropriate to this scientist's conventional notions of research design, the novelty of Theory of Change compared to conventional research planning does not lie in differing degrees of realism. By beginning with this endpoint, Lesley did not suppose that our vision represented a point that we would ever actually reach. The power of this vision lies in the horizon that it generates. Hirokazu Miyazaki (2003, 261) has highlighted how the gap between reality and ideal that characterizes utopianism generates a sense of incompleteness that "gives the present moment a future orientation" and sustains a "prospective momentum" (also see Watanabe 2019). Similarly, by asking workshop participants to begin by crafting a vision for the world, Lesley gave the subsequent work of designing research activities a new aspirational future horizon to work toward—and with it a new kind of prospective momentum.

With the aspirational future of our vision in mind, Lesley asked us to decide what changes in next-user knowledge, attitudes, and skills would be required to achieve these changes—the outcomes we wish to generate. One such outcome was: "Public policy actors develop sustainable and equitable policies (targeted toward food security, wealth, health, education, cohesion, and equity) and direct the necessary human resources and capital toward implementing those policies."

Finally, Lesley asked the workshop participants to discuss activities and outputs that could bring about these outcomes. The process of manufacturing this bottom level to the Theory of Change flowchart seems quite detached from the utopianism of the vision with which we had started. Lesley intended discussion at this moment to be pragmatic and grounded in what was feasible in the short-term future: "How could we get local government to adopt policies promoting sustainable rubber cultivation?"; "What kind of research would be necessary for stakeholders to know which agricultural systems are most sustainable?"; "Could we set up a multistakeholder platform?"; "Whom should we invite to participate?" Working backward from our aspirational future, nevertheless, meant that workshop participants sustained a prospective momentum: plans for research activities were made with a vision already in mind for the future these activities would bring closer to being.

At the time of the workshop, most of IFF's ongoing projects in the region had been initiated with no thought to outcomes or impact as these concepts are understood in outcomes thinking. Lesley asked that we redesign or retrofit existing projects so that they could be brought in line with the new Theory of Change we had just created. To fit ongoing projects into this flowchart, IFF scientists would have to generate a kind of prospective momentum that their projects had not previously required. This involved IFF researchers hastily conceiving social science research that they hoped might highlight the relevance of their ongoing biophysical research to the achievement of sustainable and equitable policies. Other IFF staff proposed workshops with stakeholder groups to share research findings in a more proactive manner than the traditional policy paper. Here, scientists were imagining a new kind of value to their work, their contribution to the construction of the aspirational future atop IFF's Theory of Change flowchart.

Theory of Change 2: M&E in the Future Perfect

Lesley's workshop focused primarily on Theory of Change as a planning tool, but many research and development institutions have also integrated Theory of Change into monitoring and evaluation (M&E) frameworks. Here Theory of Change serves as a basis for measuring and quantifying success. Scientists at IFF were already all too familiar with quantitative assessment of research through the quantification of peer-reviewed publications—either by their volume or by their quality via proxies such as impact factor or h-index. It was success defined by such metrics that had led Yin to anticipate congratulations at IFF's annual global meeting. As it was, however, IFF's global leadership had shifted its emphasis away from concern with outputs such as publications and toward success defined by the achievement of impacts. Theory of Change can operate as a basis for evaluating science against these new priorities.

The donor for IFF's Agroforestry for Myanmar project, the Food Security Fund, demanded compliance with a particularly stringent M&E framework. This framework, "Monitoring and Evaluation for Learning and Accountability" (MELA), consists of numerous components. Though they share a common grounding in Theory of Change, the contrast between Lesley's outcomes thinking and the MELA framework reveals a difference between *provisionally* anticipated effects compared to targets that are *predetermined and inflexible*. Subsumed within this M&E regime, a Theory of Change's horizon is transformed: the *aspirational future* of a world to be built gives way to the *future perfect* of quantitative targets.

A core component of MELA was a Theory of Change flowchart similar in style to the one participants had developed during Lesley's workshop. A key outcome in the new Theory of Change IFF produced for the MELA framework was "Relevant stakeholders understand better how agroforestry improves livelihoods, resilience, and ecological health in the Uplands." Working toward this outcome meant that, alongside the kind of on-farm agroforestry trials that IFF scientists would conventionally have planned, IFF scientists were having to pursue numerous knowledge dissemination activities. These included plans for hosting multistakeholder meetings and for producing an agroforestry curriculum for use by a local university.

A second component was a measurement template that demands evidence IFF is producing outcomes in the manner set out in the Theory of Change flowchart. This entailed an Excel spreadsheet with the outcomes from the flowchart listed in the first column and subsequent columns then detailing indicators and targets. So, for example, to measure the outcome "Relevant stakeholders understand better how agroforestry improves livelihoods, resilience, and ecological health in the Uplands," the IFF project team devised the indicator "Percentage of knowledge platform members with improved agroforestry knowledge" and set themselves a target of a 20 percent improvement in agroforestry knowledge by 2018.

What made the MELA framework especially troublesome for IFF scientists was not merely the intensity of the Food Security Fund's scrutiny but that MELA entailed developing precise quantitative performance indicators in advance of activities commencing. This advance determination gave detailed structure and form to a decisive retrospective moment of evaluation that awaits the project's conclusion. The MELA framework thus invited IFF staff to think in the future perfect tense: "By project completion, we will have caused X to have happened."

The MELA framework was a source of great frustration for IFF colleagues, who expended significant energy developing and revising it. One headache for the project team was the incongruity between the short-term scope of the M&E framework—which covered only the four years of Food Security Fund financed work from 2014 to 2018—compared with goals that would take many years to achieve. To address this, the project team's initial draft of the MELA framework included the measurement of indicators beyond the 2018 conclusion of the project. The Food Security Fund, however, rejected these proposed long-term targets, insisting that IFF base its measurement framework exclusively on what is achievable within the 2014–2018 funding period. Food Security Fund project managers conceded that this time frame would make achieving end-user impacts such as increased local incomes unlikely but suggested that IFF could nevertheless demonstrate next-user outcomes such as improvements in agroforestry knowledge within four years. In effect, this meant truncating the horizon of the project's Theory of Change. With impacts unmeasured, end-users became irrelevant to project evaluation. What counted were the next-user outcomes. Here, the distant aspirational future that characterized Lesley's version of Theory of Change gave way to focus on the date-punctuated near future (Guyer 2007, 417) of the donor's project evaluation.

As described previously, IFF scientists encountered an incongruity between the short-term horizon of the MELA framework and the project's long-term goals. IFF colleagues also highlighted a more profound absurdity in M&E. Jacob, a veteran of numerous agro-environmental research for development projects, told me that he finds not only the MELA framework but monitoring frameworks in general to be an "utter pain in the backside." At the root of his frustration was the demand to determine the precise course a project will take in advance of its inception. Jacob pointed out that this is an utterly unrealistic demand: one cannot know what to expect of a project before it has even begun and should rather adapt to the situation as it evolves. Then, Jacob told me, once your project is complete, you must waste time justifying why you have not met goals that you only set because you were forced to do so. From Jacob's perspective, the development of M&E metrics in advance of the project demands an inappropriate predetermination of project goals, making a farce of the ultimate retrospective evaluation these goals are used to conduct. More appropriate from Jacob's point of view would be provisional goals that can be adapted and revised as a project progresses.[3]

Though erased right from the start by the MELA framework, a provisional mode of planning is not foreign to Theory of Change. Indeed, during Lesley's workshop, participants were told to treat their Theory of Change as a "working hypothesis"—something to regularly revisit and revise. Even the MELA framework asked project implementers such as IFF to "facilitate the use of M&E results to improve the project," and in discussions of MELA, Food Security Fund project managers would often reassure IFF that they are flexible and willing to consider revisions to the project's MELA framework. Many versions of Theory of Change and M&E beyond IFF foreground precisely this kind of provisional thinking. In the context of global mental health, for instance, Dörte Bemme (2019, 585) suggests that Theory of Change operates as a "mode of knowledge production that . . . embrace[s] the epistemic instability of 'mental health' knowledge." And, in the context of development aid, Casper Bruun Jensen and Brit Ross Winthereik (2013, 148–49) observe that "perpetual revisability [is] an important dimension of the monitoring movement." IFF's experience of drawn-out interactions with Food Security Fund project managers to approve the initial version of the MELA framework, however, meant that IFF staff had little confidence in assertions of flexibility and revisability. As such, IFF staff approached the MELA framework not as a provisional position but as a fixed, future perfect image of the project's endpoint.

The rendering of MELA as offering a predetermined future perfect endpoint also reflects a tension inherent in the MELA framework's marriage of Theory of Change with a particular form of auditable metrics. As Sam—a London-based outcomes thinking consultant unconnected to IFF—put it in response to my description of IFF's experience with the MELA framework, the demand for quantifiable evidence to prove impact inevitably leads to the "bastardization" of Theory of Change as "a thought process." In Sam's view, the use of Theory of Change as a way of demonstrating impact or providing definitive assessments of success is incompatible with the conceptualization of Theory of Change as a reflexive planning process. M&E, he told me, should be used as a learning tool, as something that allows project managers to learn what is and is not working, allowing them to revise activities and objectives as they go. When I shared an earlier version of this chapter with Lesley, she offered a similar perspective. As she put it, "monitoring and evaluation of a Theory of Change is longitudinal, it is continuous and happens in real-time. The purpose is to identify when the whole effort may be drifting away from [the Theory of Change's] visionary end state, rather than moving towards it." For Sam and for Lesley, this kind of learning is inevitably frustrated by the kind of process IFF went through with MELA. The predetermined, future perfect endpoint that MELA demanded erased the potential for a Theory of Change to operate as a provisional plan subject to real-time evaluation.

This erasure is at the heart of Jacob's frustration with M&E, but it is also something that frustrates people such as Sam and Lesley for whom this erasure undermines fundamental rationales for Theory of Change and the adaptive learning goals of M&E. Despite its global reach, Theory of Change is not a singular and immutable form. The aspirational future horizon of Lesley's outcomes thinking is very different from the future perfect horizon generated by the Food Security Fund's MELA framework.

Theory of Change 3: The Eternal Impetus of Brand Building

Not long after Lesley's workshop, IFF-China hosted an IFF policy advisor and communications consultant named Alistair who promised to help the China office develop a new communications strategy. In the same seminar room where Lesley had delivered her two-day workshop, Alistair gave a wide-ranging presentation in which he extolled the lessons that science communication had to learn from the world of corporate brand management. Like Lesley, Alistair elaborated a version of Theory of Change that would empower IFF to generate outcomes and impacts. The motivating horizon, however, once again shifts: IFF's activities are to be driven not by an *aspirational future* or by *future perfect* goals but by the *eternal impetus* for brand growth.

In Alistair's view, "Theory of Change is simply another way of saying business plan." Developing this analogy, he suggested to China-based colleagues that IFF's mission to change the knowledge, attitudes, and skills of next-users is much the same as the work a business does to sell its product. What IFF needed, Alistair continued, was to get better at "selling our own product": agroforestry. To do so, Alistair proposed instrumentalizing M&E. More monitoring and evaluation of IFF's work, Alistair explained, would generate "rigorous data" on just how effective a land-use strategy agroforestry is and would underline the "value for money that we offer." As such, Alistair celebrated the fact that IFF headquarters had recently recruited a new M&E team, and he urged IFF-China to work closely with this team. Doing so, he predicted, would significantly enhance IFF-China's ability to persuade policymakers and donors of the efficacy of agroforestry as well as of the work that IFF does in researching and promoting it.

Alistair's focus on ideas like value for money echoes wider trends toward efficiency and the hypervaluation of quantitative data that I return to in chapters 5, 6 and 7. A distinctive feature of Alistair's peculiar take on marketing/Theory of Change, however, was the imperative he saw for IFF to build "agroforestry's brand." Whereas the IFF team's concerns within the MELA M&E framework focus almost exclusively on the individual project at hand, the brand-building strategy that Alistair advocates implies a much broader understanding of IFF's relationship to next-users. In Alistair's conception, the brand of agroforestry stands on a different level than the reputation of any individual project. The reputation of a well-developed brand is more than simply the sum of its products and will sustain itself despite individual instances of project failure—just as, in Alistair's example, the brand Coca-Cola survived the notorious failure of its product New Coke. Individual projects can serve as success stories with which next-users can be persuaded of the benefits of agroforestry, but this must be understood as part of a broader brand-management strategy. Projects like Agro-forestry for Myanmar are thus subsumed as means to a grander project. Alistair told the seminar audience that IFF is already reaping the rewards of his and col-leagues' ongoing efforts to build agroforestry's brand in Europe. It was because of agroforestry's growing brand recognition, he explained, that he could get IFF's director-general into an important intergovernmental forum on global security and that colleagues at another agroforestry organization had succeeded in per-suading the EU to integrate agroforestry practices into the common agricultural policy (cf. Mosquera-Losada et al. 2018). Thanks to "agroforestry's strong brand," Alistair boasted, the concept of agroforestry is now "percolating policy docu-ments." Proactive brand management had, in Alistair's view, transformed agroforestry science from an obscure body of knowledge to a vital component of multinational agricultural and environmental policy.

The way in which Alistair imagined and celebrated branding as a strategy for knowledge circulation and policy transformation shares a certain affinity with aspects of early ANT scholarship.[4] In his groundbreaking essay on the scallops of Saint-Brieuc Bay, Michel Callon (1984) describes the efforts of three marine biolo-gists to enroll scallops, fishermen, and scientific colleagues to the service of their conservation strategy. The tactics these marine biologists employed—"seduction, transaction, and consent without discussion" (Callon 1984, 214)—differ from Alistair's instrumentalization of M&E. In a more fundamental sense, how-ever, Alistair's take on scientific practice looks at lot like that of Callon's marine biologists: the goal, in its crudest terms, is for "researchers to impose themselves and their definition of the situation on others" (Callon 1984, 196). Though a pro-gram for action rather than a sociological framework, there is likewise an uncanny resemblance between the logic of Alistair's Theory of Change and Bruno Latour's (1988, 172) description of knowledge production as a process in which "[a]s soon as one actant manages to persuade others to fall into line, it thereby increases its strength and becomes stronger than those it aligned and convinced." This some-what Machiavellian reading of science, as Marianne de Laet and Annemarie Mol (2000, 227) put it, "says that technologies depend on a power-seeking strategist who, given a laboratory, plots to change the world."

Aside from his more unabashedly world-making ambitions, there is also a more subtle distinction between Alistair's rendering of Theory of Change com-pared with Lesley's or the Food Security Fund's. Whereas MELA's future per-fect horizon demands that IFF evinces its ability to effect changes within the limited time frame of a project, branding implies a collective agency that proj-ects itself into an endless future. This endless quality superficially echoes the pro-spective momentum generated by the aspiration future of Lesley's Theory of Change. Branding, however, implies a distinctive impetus. Alistair did not imagine an ideal state for agroforestry's brand: there was no generative gap between the reality of agroforestry's brand and an ideal vision of it to which Alistair aspired. Rather, an entailment of branding as a commercial strategy is the infinite imperative for growth. As a leading brand-management textbook puts it, "Brand management over time is the permanent pursuit of growth" (Kapferer 2012, 219). This imperative to pursue permanent growth gives brand building an eternal quality. Whereas Lesley's outcomes thinking generates momentum toward an aspirational future, Alistair's branding strategy implies an eternal impetus to the task of enlarging IFF's world-making power.

Theory of Change as Comparative Framework

The role that Matt had envisaged for me at IFF was centered on putting the new way of thinking Lesley gave us into practice. One of the outcomes-orientated activities colleagues and I in Matt's research group devised was an experiment—conducted in collaboration with a county-level forestry bureau and local mushroom harvesters—to compare a range of indigenous mushroom management practices. We conceived this project as a means to build on IFF's extensive existing mycological research and expertise in a manner that would satisfy the imperative to generate outcomes and impacts. Our target impact was for mushroom harvesters to adopt more productive, and therefore more profitable, mushroom harvesting practices. Working backward from this end-user impact, we had formulated a next-user outcome of forestry extension workers gaining improved knowledge of mushroom management. We therefore decided to begin by devising an experiment that would produce new knowledge of mushroom management that we could pass on to forestry extension offices.

Not only did Theory of Change provoke IFF scientists to conceive of this new research project, but it also gave us a new means for evaluating a project's success—or, in this case, failure. Having completed the mushroom management experiment, data analysis showed no significant difference between the various management strategies that we had compared. Reporting these results to Matt, I suggested that although our research could not say anything of practical use to mushroom harvesters, our data might have the potential to generate a decent scientific paper. Matt laughed at this. By rendering a peer-reviewed paper as the project's only output, I had forgotten the point of the research. We had originally anticipated that our comparative analysis would allow us to propose a best practice that we could subsequently disseminate to harvesters (the end-users) via forestry bureau extension officers (the next-users). Our goal had been to generate outcomes and impacts, and publishing research results could not, as I had implied, stand in for this goal. In this respect, our project was somewhat of a failure.

There is also a sense, however, in which by arriving at this conclusion both Matt and I had adopted a distorted version of Lesley's Theory of Change. In our impatience to satisfy the imperative for impacts, we had hastily imagined a Theory of Change that forgot the place of researchers as next-users. Had we done so, we might have considered how our research (even with null results) might have helped us persuade scientific colleagues of which further avenues for research would and would not be worth pursuing. This would have involved considering how shaping the knowledge and attitudes of scientific colleagues might serve as a causal link in a more complex pathway to our target impact of enhancing sustainable mushroom yields. As Lesley highlighted in response to an earlier version of this chapter, moreover, rather than seeing these null results as a failure of the project, we might have taken the deviation from our intended impact pathway as an opportunity for "real-time evaluation." Seeing the project drift away from our Theory of Change's causal pathway, we might have taken the opportunity to rethink the pathway to our aspirational future. Though we had taken on board many of its fundamental lessons, we had perhaps also transformed Theory of Change into something that was narrower and more rigid than what Lesley described in her workshop.

When he laughed at my appeal to publication as the project's endpoint, Matt had nevertheless taken on board a fundamental lesson from Lesley's workshop. The short shrift Matt gave to my proposal for a publication as the project's endpoint highlights the novelty of the structure that outcomes thinking imposes on IFF research. The near-future endpoint of a peer-reviewed article is no longer an acceptable horizon to work toward. Driven by the gap between reality and the aspirational future of IFF's vision for the world, Lesley's Theory of Change forces scientists to focus instead on a protracted causal chain in which their activities will produce next-user outcomes that will in turn produce end-user impacts.

Most of the participants in Lesley's workshop were, like Matt and me, encountering Theory of Change for the first time. Bob, a China-based research fellow, was an exception. Tasked with setting up the Lengshan Multi-Stakeholder Platform, Bob saw his role at IFF as being somewhat like Lesley's—one of "improving the connection between scientific research and development impacts." Reflecting on Lesley's workshop, Bob suggested to me that everyone at IFF already works according to "implicit Theories of Change." What the workshop was doing, in Bob's analysis, was simply forcing scientists to make their tacit Theories of Change explicit and to reflect on them. In a certain quite narrow sense, Bob's assertion seems incorrect: his colleagues conceptualize and practice science in a manner that is incommensurable with Lesley's Theory of Change. Notwithstanding the disjuncture between Lesley's conceptualization of Theory of Change and IFF researcher's translation of it into practice, Matt and many of Bob's colleagues recognize the radical transformation that Theory of Change implies for scientific practice. Theory of Change frameworks demanded scientists engage the future in ways that were previously unimaginable.

While IFF scientists' usual practices were very different from those prescribed by Theory of Change, there is nevertheless also a more important sense in which, following Bob, we might say IFF scientists operate according to implicit Theories of Change. Though in a different structure from the one Lesley espoused, IFF scientists likewise imagine future effects for their work—effects that are understood and pursued according to a tacit structure that, as Bob pointed out, is rarely made explicit.

Implicit Theory of Change 1: Endpoints and Unanticipated Futures

As familiar as Alistair's brand-building strategy may be to Callon's account of marine biology in Saint-Brieuc, it is incongruous with how IFF staff ordinarily conceive of scientific practice. Indeed, the world-making ambitions of Alistair's brand building jar with the more modest ambitions of most IFF research staff. Equally, IFF scientists' do not usually imagine themselves working toward the kind of aspirational future that Lesley had them formulate in her workshop. More vaguely, though no less earnestly, IFF scientists aspire to contribute to endeavors such as addressing climate change, safeguarding biodiversity, and improving rural livelihoods. Scientists' imaginations of the connection between research outputs and the actualization of such endeavors are, moreover, generally free of any pretension to a plan. Instead, IFF scientists more often focus their attention on the endpoint of a publication while fostering hopes that the knowledge such publications contain might be somehow put to an as-yet-undefined use in the hands of an as-yet-unidentified other.

This orientation was evident in a paper presenting an overview of edible mushrooms in Asia to which several IFF colleagues contributed. This paper concludes with a discussion section highlighting, among other things, how little is currently known about mushrooms of the region. As Matt, one of the coauthors, described it to me, the paper's goal was to "provoke activity" in Asian mycology. In this paper, there is closure of a kind: a desire to establish Asian mushrooms as a "matter of concern" (Latour 2004b). Control over how this concern plays out was nevertheless ceded to others (de Laet and Mol 2000). Establishing mushrooms as a problem for future scientists is, of course, no neutral act. No less than pursuing outcomes thinking or brand building, provoking activity is an act that (if successful) would compel others to action. The structure and sequencing of these anticipated effects were, however, fundamentally different from what Lesley or Alistair had suggested. Whereas a Theory of Change flowchart begins with a vision of the future IFF will compel others to build, Matt and his colleagues' mushroom paper does not preconceive the future its work would bring into being.

This incongruity was apparent in my misunderstanding of another colleague's desire to conduct applied research. Staying late at the office one evening, I struck up a conversation with a postdoc named Ruyue. When I asked her what she was working on, she launched into an excited description of a proposal she was writing to compare the carbon storage capacity of different mycorrhizal fungi. One of the reasons this research interests her, Ruyue explained, is that she wants to use her expertise in soil ecology to do "applied research." Ruyue had coauthored numerous articles in top scientific journals, but she was frustrated that those prior research activities had been so far divorced from practices of rural land-users. This would be different in her new research. IFF, Ruyue explained, already has databases capable of aiding a forester's tree selection, but these databases ignore the importance of the various soil mycorrhizal fungi with which these tree species can symbiotically coexist. Her proposed research, Ruyue continued, would provide the knowledge necessary to create a complementary database to guide selection of mycorrhizal species for inoculation into forestry and agroforestry systems.

Recalling her earlier reaction to Lesley's workshop, I was surprised by Ruyue's enthusiasm for "applied research." Ruyue had attended day one of Lesley's workshop but, along with several colleagues, had made her excuses for day two. She had also joked with me at coffee the day after that I would "get used to this sort of workshop"—a dismissive reference to the steady flow of international experts who would pass through Songlin to advocate for and train IFF-China colleagues in whatever happened to be IFF's newest strategy revision or research framework. Perhaps, I wondered, Ruyue's thinking was closer to Lesley's than she realized. Channeling what I had learned from Lesley's workshop, I responded to Ruyue's description of her new project proposal by inquiring about how she would turn her research into what Lesley would have labeled outcomes and impacts. "How do you plan," I asked, "to persuade policymakers and land-users to use your database?" Ruyue responded with a bemused laugh and suggested that I was getting ahead of myself. "First, I need to do the research," she explained. For Ruyue, worrying about the specifics of how she might engage policymakers and land-users should wait until after her research was done and the database completed.

I had been mistaken, it turned out, to assume that Ruyue's interest in applied research implied anything resembling outcomes thinking. Lesley had described Theory of Change as beginning with an aspirational future. Working backward from this vision, Lesley's Theory of Change entails working through a causal chain of impacts, outcomes, and outputs. Only once this protracted chain of future effects has been mapped out does Lesley think it appropriate to decide on research activities to pursue in the present. Encouraging IFF scientists to engage multistakeholder platforms, moreover, Lesley compels scientists to identify and cultivate relationships with next-users prior to beginning research. Ruyue, by contrast, thinks it inappropriate to make concrete plans beyond the near-future horizon of conducting research. Ruyue designs research that she hopes might be useful—that anticipates application—but she has no designs for precisely who will make use of her research. When Ruyue imagines the future of her research, she, like Matt, defers the problem of planning its concrete application and ultimate impact on the world to the future.

The inflexible determination of a Theory of Change flowchart and associated target outcomes that is found in MELA is not inevitable to Theory of Change. Theory of Change flowcharts are, like a research proposal, often imagined as merely provisional. The distinction between Ruyue's applied research and Theory of Change is therefore not merely that one is open-ended and the other is not. The difference lies instead in different forms of open-endedness. Though both a Theory of Change flowchart and a research proposal can operate as a provisional plan, a key difference is that a research plan has an actual endpoint. In a conventional research proposal, the research will be finished, and the plan will reach its conclusion. As described previously, the vision that sits atop Lesley's Theory of Change did not represent an actual destination; it was an aspirational future that imbued work toward it with prospective momentum. Alistair's brand-building strategy, meanwhile, derives its momentum from the eternal impetus for brand growth. By contrast, Ruyue and Matt imagine an actual endpoint to their research: Matt's research was already completed, and Ruyue's would be. This is an endpoint, however, that scientists hope will generate further as-yet-unknowable effects. It is at once an endpoint and the genesis of unanticipated futures.

Implicit Theory of Change 2: Public Interest as Endpoint

Though Theory of Change first arrived at IFF's China office in the wake of external admonishments that IFF-China had failed to generate outcomes, many IFF-China colleagues shared Ruyue's discontent with their individual and collective failures to contribute to social and ecological change. Trained in a botany department, Matt would often differentiate his interest and expertise in pure or fundamental science with what he saw as the more applied orientation of a research for development institute like IFF. Nevertheless, he embraced the mission of his employer and, initially at least, the opportunity that Lesley's workshop offered to learn how to turn insights from research into outcomes and impacts. In 2014, this was learning that Matt believed he and his colleagues dearly needed if they were ever going to contribute to IFF's long-standing mission to research and promote agroforestry. For all the time, money, and energy he and colleagues had invested toward that mission, Matt lamented that the most significant impact his work had made on local landscapes was a "few badly planted trees by the side of a road"—a reference to a heavily funded but poorly executed landscape restoration project.

Another colleague, Jiaolong, often joked that the inevitable conclusion from my ethnographic research would be that "IFF is useless." For Jiaolong, the failure of her and her colleagues to have any positive effect on the world is a frequent preoccupation and is a problem that requires scientists to rethink how we engage the world. As she put it in one of her blogs, "Scientists often complain that the public don't understand science and don't think that popularizing science is important, but the question is, have we helped to foster their interest in science?"

The inspiration for the blog, Jiaolong explained over a Friday evening beer, was a workshop she had attended with other recent PhD graduates in the Netherlands. Arriving at the first day of the workshop, Jiaolong introduced herself to one of her peers, telling him that she had really enjoyed reading his doctoral dissertation. His response took her aback. Laughing, he asked her, "Why on earth would you want to read my dissertation?" This flippant response referenced the fact, of which Jiaolong was well aware, that very few dissertations are read by more than a handful of people. To Jiaolong, her peer's attempted joke begged a serious question: "Why on earth are we producing all this research if we know no one is going to read it?" For Jiaolong, this is a question we should be asking not just of doctoral dissertations but of academic publications in general. "Even when someone publishes in a top journal," Jiaolong asked me rhetorically, "how many people are really reading that work?" Indeed, as she points out in her blog, most scientific publications are not only hidden behind a paywall, but they are also published in a language that most Chinese people do not read: English. Rather than continue aimlessly churning out dissertations and journal articles, Jiaolong's blog urges us to ask: "Can we use different methods to more quickly transmit our research results to our primary and middle school classmates, to our classmates' classmates, to our kin's kin?"

Jiaolong's reference to classmates and kin evokes her own personal relationship to the region in which IFF works. She had grown up in Lengshan and many of her former classmates and most of her kin work on the very farms and forests that she and her colleagues examine in their research. When she began a PhD investigating rubber plantations, she had been driven by the idea that her research might help address the ecological and economic challenges she sees in her own community. Her joke about IFF being useless, in this respect, reflects the frustration of a long-standing aspiration. Nevertheless, Jiaolong maintained an enduring faith in the possibility of using science to do good in the world. To be sure, Jiaolong did not pretend to have an easy solution to the problem of making science generate positive change, but she does demonstrate the possibility for making this a very different kind of problem to that posed by Lesley, the Food Security Fund, or Alistair.

Taking Jiaolong's imperative to communicate our research to classmates and kin, we might ask: What did Matt and the coauthors of his mushroom paper do to foster interest among publics beyond professional scientists? This question would draw attention to the fact that when Matt told me he wanted to "provoke activity," what he meant was activity among other scientists rather than among any broader public. Though Jiaolong's question highlights a failing in Matt's work, it is a very different failing than the failings that Lesley's outcomes thinking, the Food Security Fund's M&E framework, or Alistair's branding strategy would highlight. Rather than asking what Matt did to foster public interest, evaluating his research within these Theory of Change frameworks might lead us to ask: How is Matt's paper bringing us closer to the vision of the world we want to bring into being? What changes in the knowledge, attitudes, and skills of specified next-users was his paper meant to effect? What measurable outcomes has his paper generated? Or what does his research do to enhance agroforestry's brand? Each of these questions implies quite a distinctive imagination of how science generates effects in the world.

Jiaolong's proposal is to shift the endpoint of scientific practice from the point of presentation to scientific colleagues—publication in peer-reviewed journals—to the point of presentation to a broader audience. Like Lesley's outcomes thinking, Jiaolong draws attention to the limits of peer-reviewed (and peer-read) journals as the primary horizon for scientific practice. Compared with Lesley's approach of identifying specific next-users, however, Jiaolong's call to interest our classmates' classmates and our kin's kin suggests a somewhat amorphous public—a notion that echoes the similarly amorphous scientific community among whom Matt wished to provoke activity. Nevertheless, whatever the audience or medium for Jiaolong's public-oriented endpoint, she offers us a very different kind of challenge than the one implied by Lesley. In Jiaolong's implicit

Theory of Change, the impetus for scientific practice to transform the world does not depend on the prior elaboration (provisional or otherwise) of an aspirational vision for the future. Jiaolong (or, for that matter, Ruyue and Matt) is no less concerned than Lesley with the desire for IFF's research to be useful. The difference, however, is in the sequencing and structure of expectations for utility: when, how, and by whom research is put to use as well as when, how, and by whom this use is anticipated and planned for.

The Theory of Change that Lesley asked her colleagues to embrace implies a transformation in the structure of scientific practice. Rather than allowing scientists to cede the problem of how research is used, Lesley asks that scientists plan for how, when, and by whom their research findings will be put to use. The impetus for these designs is a new kind of horizon, an aspirational future world that research strives to bring into being. This is a very particular way of imbuing the present with a future orientation—one in which a gap between a reality and a projected ideal would sustain prospective momentum. Diverse iterations of Theory of Change, however, demonstrate the multiplicity of future orientations that can structure knowledge production. In this regard, the aspirational future that orients Lesley's Theory of Change implies a very different orientation than does the future perfect deliverables of MELA. This is again different to the futurity implied by the eternal impetus of brand growth contained in Alistair's rendering of Theory of Change as a business strategy. The diversity of ways in which this one planning and evaluation technology is operationalized within the same institution suggests that demands for impact are ubiquitous but by no means uniform.

At issue in the divergence between the iterations of Theory of Change—as well as between them and colleagues' implicit Theories of Change—is not merely whether science is expected to generate practical or concrete effects in the world but a multitude of distinct and often mutually incompatible ways of structuring and motivating scientific practice. Despite their ostensible common grounding in Theory of Change, Lesley's outcomes thinking, the Food Security Fund's M&E, and Alistair's branding strategy each suggest very different models for how scientists engage stakeholders beyond the academy and for how we imagine the role of scientists in shaping the future. Jiaolong's call for her colleagues to foster public interest in the sciences, moreover, demonstrates that highlighting a lack of vision, the absence of measurable targets, or the inadequacy of a scientific brand are not the only extant challenges to contemporary science nor the only starting points for reimagining its future.

More so than the heterogeneity of models for impactful science that animate the audit cultures of IFF and its donors, however, I want to underline the provocation that Theory of Change provides for us to reflect on the structure of scientific practice as it is and as it might become. Again, this is not simply a question of a straightforward difference between a theoretical and an applied orientation. Even before Lesley's workshop, many IFF colleagues—much like many of us in the humanities and social sciences—already aspired for their research to make a positive difference in the world. The value that Bob located in Theory of Change, however, was in its capacity to render explicit scientists' existing aspirations and to thereby create an opportunity for critical reflection on them. As I have shown, Lesley introduced IFF-China to a Theory of Change that entails distinctive imagination of the horizons toward which scientific research is to be practiced, the points in time at which the future effects of research are to be anticipated, and how scientists' agency is to play out across time. Following through on Bob's suggestion that IFF scientists already had implicit Theories of Change has meant elaborating IFF scientists' aspirations and practices in terms of these same dimensions. Doing so not only allows an appreciation of how diverse iterations of Theory of Change might differ from prevalent models of scientific practice; it can also help us appreciate latent alternatives to the peculiar structures of impact demanded by contemporary audit cultures. In this regard, we can do more than simply critique the logics and sensibilities of planning and evaluation tools like Theory of Change; we can also borrow and repurpose them to do what I have begun to do in this chapter: to make explicit the structures—the implicit Theories of Change—of IFF colleagues' existing knowledge practices. The chapters that follow continue to borrow Theory of Change's orientation to temporality and agency as a framework for understanding scientific and bureaucratic practices at IFF as well as for reflecting on the task of remaking scientific practice.

BUREAUCRACY AS INTERRUPTION

Making small talk at an office lunch in honor of a visiting delegation, Chang-kong, a PhD student in forest ecology, asked me about my research. I explained to Changkong that I was interested in how Chinese government regulation restricts IFF's research. Changkong gave me the same bemused look that several of his colleagues had earlier given me when I had framed my research in this manner. I quickly corrected myself, explaining to Changkong that I was interested in the "*mafan*" that accompany scientific research. Changkong responded with a knowing smile and added that he had been lucky to encounter far fewer *mafan* than many of his classmates.

The term I used here to frame my research interest is a crucial part of IFF colleagues' lexicon. Even international staff who spoke little or no Chinese would seamlessly throw the term into an English sentence: "Is this going to create a *mafan*?"; "This sounds like a lot of *mafan*." When staff at IFF speak in English, they use *mafan* interchangeably with "headache." In the interests of readability for non-Chinese speakers, I will refer to this concept with the English term "headache." In this chapter, I elaborate on the meaning and significance of headaches within the scientific, administrative, and bureaucratic lives of IFF colleagues. I also begin to develop the sense of interruption inherent in headaches as a framework that might help us reimagine the relationship between science, bureaucracy, and society.

By the time of my conversation with Changkong, I was already several months into my two years at IFF. The appreciation I had of headaches was the product of numerous prior failings to communicate my research interests to colleagues.

Having taken to the field an interest in regulatory and bureaucratic obstacles that IFF faced, I would often explain to IFF colleagues that I was interested in how IFF's work was "restricted" or "controlled" by state regulation. I was at first surprised to hear the response that IFF's work is in no way restricted. This was a claim that seemed utterly incongruous with the ubiquitous bureaucratic headaches that I had witnessed. Unable to understand, I would describe to IFF colleagues instances of what I meant by government-imposed restrictions. I would point out, for example, that IFF staff must apply for an official letter of introduction to give to local government whenever they leave campus to conduct field research. And I would point out that even with such a letter, government officials were not always willing to sanction IFF's activities. In an instance that I often offered as an example, a graduate student named Feixue wanted to carry out a survey of macrofungi inside a Chinese government reserve. In this case, the reserve administration initially refused to accommodate Feixue, and it was only after a protracted negotiation that she was able to belatedly embark on her data collection. When I described the difficulties encountered by Feixue to IFF staff, however, they would respond that that was an example not of restriction but merely of a headache.

My suggestion of restriction had implied that regulation and bureaucracy prevent activities from happening, whereas the examples I had observed represented only temporary difficulties: mere interruptions. Mubai, an IFF PhD student and administrative assistant, likened these headaches to mosquitoes: "You can easily squash them," he told me. He also invoked a popular idiom: "Those in power have their policies, those below have their ways of getting around them" (*Shang you zhengce, xia you duice*). My mistake in discussing restrictions had not been to see intense bureaucratic work where there was none. It was rather to have misread the temporal character of bureaucratic work. For Mubai and his colleagues, the bureaucratic challenges that they routinely face do not alter the work that IFF ultimately does. When IFF staff confront bureaucratic challenges, they are confident that they will overcome the obstacles in their way. It would take significant labor to get Feixue access to her research site, but IFF staff had always had faith that ultimately Feixue's research would proceed as planned. As IFF staff understand it, a headache interrupts work but does not restrict or transform it.

IFF colleagues' apprehension of bureaucracy as a headache is something that also extends to the administrative work of research funding and of academic careers. One particularly persistent headache for IFF staff is the MELA framework discussed in the previous chapter. For the IFF scientists involved, Agroforestry for Myanmar provided an invaluable opportunity for new and important research on agroforestry. This was research, moreover, that staff hoped might be put to practical use by upland communities in Myanmar. In a certain sense, this goal was shared by the Food Security Fund and promoted by its MELA framework. As described in the previous chapter, however, IFF staff lament the overdetermined nature of the targets demanded by MELA and treat its evaluation framework as a distraction from what they see as the real work of producing and disseminating scientific knowledge. Labeling MELA a headache—or, as Jacob put it, "a pain in the backside"—the IFF project team approached the MELA framework in a manner similar to Feixue's encounter with state bureaucracy. Satisfying the Food Security Fund's bureaucratic demands was an interruption to be dealt with so that the real work could continue. Whereas advocates of apparatuses such as Theory of Change celebrate its ability to generate *momentum* toward impactful research, for Jacob and his colleagues, the only thing MELA's Theory of Change generated was meaningless *interruption*.

A headache, in these instances, was meaningless in two senses. First, a headache, in contrast to a restriction, is of little consequence to the work that it interrupts. IFF staff are animated by a faith that headaches are manageable—that they waste time but do not transform the work that they interrupt. This faith is sometimes strained by the intractable and relentless nature of certain bureaucratic and administrative difficulties, but it is nevertheless a faith that underlies IFF staff's engagements with bureaucracy. Second, headaches are meaningless in the sense that, from IFF colleagues' point of view, headaches have no intrinsic meaning or value. This meaninglessness stands in contrast to the intrinsically valuable research and development work that a headache interrupts—work that awaits on the horizon once the headache is overcome.

The Temporalities of Headaches

The future horizons that IFF staff imagine beyond a headache are closely tied to the ways that staff members find meaning in IFF's work as well as to how they imagine their position at IFF in relation to personal career trajectories. Ruyue, for example, is nominally a research fellow at IFF, but most of her time is occupied by compiling project reports for a Chinese Ministry of Science and Technology (MOST) grant that supports the Asian Fungi project. In contrast to MELA, funding reports made to MOST do not follow a strict formula. MOST nevertheless expects IFF to submit reports that describe how funding is being used and that evince delivery of work promised in the original project proposal. During an interview, Ruyue described to me how it was common for funds from MOST grants to take up to two years to reach awardees such as IFF. This caused major headaches for project reporting because, even if no research funds would arrive until the second year of the nominal funding period, IFF was expected to submit an annual report to MOST at the end of the first. In short, IFF had to report activities before it received the cash to complete them. One way that Ruyue and her colleagues dealt with this was to report research that had been completed as part of entirely separate projects. Here, ostensibly unconnected research activities were reported to MOST as if they had been funded by MOST. Specifically, Ruyue reported research on crop–fungi symbiosis that a colleague had completed prior to the MOST grant being awarded and with funds entirely independent of it. As described in chapter 1, IFF scientists' responses to MELA are suggestive of an absurdity inherent in the framework's demand for predetermined outcomes. Here, Ruyue locates a different but no less absurd temporal effect in the demands of MOST project reports. Whereas the MELA framework frustrates because it generates overdetermined goals, the greatest source of headaches in MOST project reporting is the pressure to produce outputs prior to receiving funds. The experiences of overcoming the headaches of MOST reporting and of MELA are nevertheless similarly confronted as unwelcome and meaningless interruptions.

Ruyue attributed being lumbered with this reporting work to her relatively junior position at IFF as well as to her failure to win research funding of her own. She compared her situation with that of a colleague who had recently won a Natural Science Foundation of China (NSFC) grant. This grant, Ruyue explained to me, freed her colleague from the administrative duties that dominate Ruyue's professional life and allowed this colleague to pursue research. Ruyue continued to apply for her own funding, but in the meantime, her role within IFF remained largely administrative. Vital here is this work's in-the-meantime character. Ruyue's work overlapped heavily with IFF's corporate services team, but whereas many corporate services staff consider report writing and grant administration to be a core aspect of their professional work, Ruyue imagines this report writing a temporary phase in her scientific career. Indeed, Ruyue remains animated by a passion for a future research project, the outline of which she already has in mind.

On one occasion, having concluded a thirty-minute interview that explored Ruyue's work writing funding reports on behalf of senior scientists, I returned to my desk in IFF's open-plan office. Unsatisfied to have discussed only the unfulfilling aspects of her professional life, Ruyue wandered over to my desk to continue our discussion, only this time focused on her proposed research. Speaking excitedly, Ruyue described her ambition to embark on research comparing the carbon storage capacity of mycorrhizal fungi in agroforestry systems. As discussed in the previous chapter, Ruyue hoped that this research would not only demonstrate important ecological benefits of tree–fungi symbiosis but also give land-users and policymakers the knowledge necessary to select tree–mycorrhiza combinations that maximize a forest or an agroforest's carbon storage potential.

It is the promise of subsequently pursuing this research that sustains Ruyue's perseverance through the tiresome administrative burdens of her current position.

Research That Evaporates

This anticipation of meaningful work waiting on the horizon echoes Matt's experience of navigating the MELA framework—something he is responsible for as Agroforestry for Myanmar's project leader. As it is for Ruyue, Matt's in-the-meantime work of overcoming headaches is animated by a faith that present headaches would eventually give way to scientific research. His commitment to research is sustained because work in the meantime, though entirely tangential to research itself, is imagined as merely a temporary interruption.

The labor of navigating these headaches, however, is unevenly distributed. Specifically, Matt delegates the MELA headache to junior colleagues. Shortly after the Food Security Fund introduced the MELA framework to the project, Matt recruited an additional staff member, Kathryn, into his project team. With her experience working with NGOs and her expertise in the workings of international donor organizations like the Food Security Fund, Kathryn brought to the team a familiarity with M&E methods. This experience and expertise, Matt hoped, would allow Kathryn to deal with the headaches MELA brings so that he and his research team could get on with their research.

The horizons and duration of the headaches that Matt and Ruyue experience diverge beyond the mere fact of Matt's ability to delegate administrative work to more junior staff. If Kathryn can deal with MELA, Matt can return to a research project that exists in the present. For Ruyue, by contrast, the horizon of return to research is more distant. The research she dreams of is something that only a research grant or fellowship would make possible, and Ruyue could not be sure when—or if—this grant would arrive. The distance of this horizon, however, does not mean that Ruyue's view of the future is less optimistic than Matt's. For Ruyue, as for many junior academics, she views her position as a stepping-stone to something better: to a later phase of her career when she anticipated having the funds, the position, and the time to conduct exciting new research. Matt's perspective is almost the reverse. For him, his seniority seemed to have robbed him of his scientific career. Having first-authored numerous papers based on his PhD research, after five years at IFF, Matt's research output was increasingly limited to smaller contributions as a coauthor. This is something that Matt attributed to his responsibilities supervising students and leading a research team, but most importantly he blamed the proliferation of administrative work that came with his newfound seniority. Walking to lunch one day discussing the latest headache that MELA had thrown up, Matt lamented to me, "I remember when I used to get to think about research." In this moment, headaches had proliferated to the extent that they ceased to be interruptions and had instead become the very essence of Matt's professional life. Ruyue imagines herself on a trajectory toward having more time for science—a time when headaches would dissipate. Matt, by contrast, fears that he might be on a trajectory toward less and less scientific work—toward a time when he will effectively become a full-time administrator (cf. Brenneis 1994, 33). The intractability of the MELA framework challenged Matt's faith in a headache's in-the-meantime character. In this moment of crisis, Matt encountered his return to meaningful research as an evaporating horizon.

A Collective Mission

Matt's fear that he might be becoming an administrator points also to the divergence between how bureaucratic headaches are experienced by scientific staff compared with corporate services staff, many of whose professional work is already defined by its administrative domain. The bureaucratic travails of the corporate services team who occupied the office's third floor were well known—and usually well appreciated—among the research staff on the floor below. One corporate services colleague who has a particularly notorious workload is known affectionately as *Biaomei*—a play on words that translates literally as "form sister."[1] This colleague's work, unlike Ruyue and Matt's, is not animated by the promise of returning to her own research. Nevertheless, corporate services staff's energies are often animated by a commitment to IFF's research for development activities. Some colleagues, for example, told me that they had taken their jobs with IFF because they wanted to support IFF's research and its aspirations for environmentally sustainable rural development. For these staff members, bureaucratic headaches sit in the path not of personal academic careers but of a collective institutional mission.

This is not to say that corporate services staff have a wholehearted belief in IFF's achievements. Much like their scientific colleagues, corporate services staff on occasion express frustration with IFF's inability to do anything worthwhile. Echoing Jiaolong's joke that IFF is useless, Jianming, a senior member of the corporate services team, described to me how he had come to realize that in all the years that he had worked there, IFF had done nothing of value. This was a frustration that Jianming framed in relation to his long-held ambition to work in rural development and the decade or so he had spent building what he once thought of as his dream career. Determined to work in this field, Jianming worked as a volunteer before finding his first development job as a field assistant for a German project in Tibet. Jianming spoke of his boredom during the many months he spent living in rural isolation and managing day-to-day project activities, but he nevertheless found the work fulfilling enough to subsequently apply for a research assistant position at IFF. In his early days at IFF, Jianming was often called on to accompany foreign scientists to the field, helping them communicate with local land-users and government officials. Later, Jianming moved to join IFF's corporate services team, first as a finance officer and then in a brief stint leading the team. By the time he had climbed the ladder to this position, however, Jianming had become disillusioned with the value of IFF and of international development more broadly.

Over a beer one evening, I asked Jianming what had led him to see IFF in such a negative light. He gave two answers. The first was his decision to pursue graduate studies in economics. These studies gave Jianming a new set of analytic tools for evaluating IFF's work. In this respect, Jianming complained that IFF has never bothered to calculate the "return on investment"—the benefit to rural communities relative to the resources IFF expends—for any of its projects.[2] Jianming had no doubt in his mind what such calculations would show: that IFF's work was "a waste of money." The second turning point that Jianming identified was the death of a close family member. This loss provoked in Jianming a period of reflection and contemplation during which he questioned the value he had long assumed to inhere in his chosen profession.

This retrospective evaluation of IFF's worthlessness and his loss of faith in IFF's mission erased the future orientation that had once animated Jianming's professional life. The meaningful actions that he had once imagined on the horizon—the foundation and motivating force to the implicit Theory of Change that undergirded his professional life—had evaporated. Where solving headaches and navigating bureaucracies once seemed worthwhile because of the collective mission that administrative work served, Jianming came to see his entire job as a waste of time. Jianming decided to leave IFF for a career in finance. If he was going to do something pointless, he reasoned, he may as well make good money doing so.

Matt, Ruyue, and Jianming each experience the bureaucratic and administrative headaches that punctuate their professional lives as interruptions. There are important specificities to how each colleague approaches their administrative and bureaucratic burdens. Depending on one's role at IFF, interruptions might be shorter or more sustained, frequent or infrequent. Efforts to overcome headaches are, moreover, animated by quite different horizons—from the promise that bureaucratic work will subside once one's academic career progresses to a belief that there is no worthwhile future to IFF's work.

Between Headache and Feedback

In his ethnography of the development profession in Tanzania, Richard Rottenburg (2009) describes the formal evaluation of development projects in terms of Latourian centers of calculation.[3] According to Rottenburg, evaluation processes render projects into immutable mobiles that can be subsequently recombined into representations in a manner that suits the interests of the Western donor community—the center of calculation. Rottenburg suggests that the donor community renders hegemonic effects invisible, masking their exercise of power over developing countries through a rhetoric of equality enacted in the process of project evaluation.[4] Extending a similar concern with visibility to the audit regimes of Euro-American universities, Chris Shore and Susan Wright (2015, 422) draw attention to "what audits and rankings bring into focus and what they render invisible or unsayable."

This focus on visual imagery perhaps reflects the ways in which evaluation and audit regimes are discussed by the development professionals and the academics that Rottenburg, Shore, and Wright encounter through their ethnographic research. As Don Brenneis (2006, 43) notes in relation to the documentary nexus of the US National Science Foundation (NSF) funding program: "A first striking feature of these documents and of the ways in which they are discussed and written about is the centrality of visual imagery. Paired terms, especially transparency/opacity and visibility/invisibility, infuse both institutional considerations of such documents and scholarly attempts to understand them." It is equally striking how marginal such concern with visibility was among the scientists and corporate services staff whom I worked alongside at IFF.[5] When IFF colleagues dealt with the headache of M&E, a project report, or a visa application, the veracity or transparency of the representations that documents produce was of little concern (cf. Cavanaugh 2016; Giri 2000; Jensen and Winthereik 2013, 93–120; Tracy 2016). What mattered was simply that the documents produced allowed staff to overcome a headache as efficiently as possible.

Lisa Bornstein (2006) describes a similar set of tactics for manipulating M&E as a routine part of NGOs' management of international development funding. Such tactics also parallel some of the ways that Shore and Wright (2015) describe academics "gaming" the audit regimes of Euro-American universities. Shore and Wright point to a study by Michael Sauder and Wendy Espeland (2009) that describes how, to boost their position in various ranking systems, US law schools carefully manage statistics on students' entrance exam scores, postgraduation work placements, and LSAT scores. This manipulation of evaluation frameworks echoes how, to satisfy donors such as MOST, IFF colleagues incorporate into project reports research that had no tangible connection to the funding in question. It bears a strong resemblance, likewise, to how IFF colleagues proposed soft and easy-to-satisfy indicators and targets for the MELA framework. Shore and Wright are critical of such gaming, arguing that they risk creating "organizational schizophrenia." Rather than simply "buffer" ourselves against the impact of audit regimes, Shore and Wright (2015, 431) argue that we should transform them: "Another potential strategy is to take back control over the measures used to evaluate professional performance by creating alternative experts and systems of evaluation and insisting that organizations be evaluated in their own terms." The imperative Shore and Wright identify is for the subjects of audit to take control of the terms of their own evaluation and to take responsibility for finding better systems for making our performance visible.[6]

Notwithstanding the shortcomings that Shore and Wright highlight of gaming as a response to audit and to evaluation frameworks, there is more to be learned from the ways that IFF staff respond to audit systems than simply that their responses are inadequate. Colleagues at IFF analyze bureaucracies—including the kinds of academic and development regimes described by Rottenburg and by Shore and Wright—not in terms of the adequacy of the representations they generate but in terms of the interruptions they cause. Shore and Wright's call to take control of the terms of evaluation is the logical response to audit understood as a problem of visibility. Understood as a problem of interruption, however, bureaucracy and audit might pose a different kind of challenge: one focused on the quality of interruptions.

In this respect, there is an important sense in which, for IFF scientists, not all interruptions are headaches, or at least not necessarily so. Whereas IFF scientists seem to inevitably experience certain activities as a headache—navigating visa applications or dealing with the Food Security Fund's MELA framework, for instance—others can be more ambiguous. A PhD defense at IFF is one example of an activity that in many instances appears a headache but that in others can become something quite different. The PhD defenses of two international students—David and Peter—created significant headaches for two of their supervisors at IFF. David and Peter were studying at IFF as part of a broader collaboration between IFF and their home country's forestry department. These two students receiving their PhDs was to be an important milestone for this ongoing collaboration. The limited time that these two students had in China, however, was not sufficient for them to be able to complete research to a PhD standard. As a deadline for these students to complete their studies and return home approached, fears mounted that they would not satisfy the external examiners for their PhD defenses. Such was the importance of these students graduating to IFF's ongoing work with these students' home country that Professor Yin asked two of the students' supervisors to write or rewrite large chunks of the dissertations on behalf of their students.

David's supervisor was a member of Matt's soil biology group named Susanna. Much like the work of procuring research permissions or completing donor project reports, David's impending PhD defenses became an unwelcome headache to be dealt with and then forgotten as soon as possible. Susanna's frustration was not that David's doctoral research was in itself without value. Susanna had worked closely with David, as well as several other IFF colleagues, investigating soil restoration on former open-pit mines. Susanna did not doubt the importance of mining restoration, and she believed that IFF's research could contribute valuably to the transformation of degraded landscapes into productive and ecologically diverse agroforestry systems. David getting through his PhD by hook or by crook, however, did not further this research agenda. Nor, for that matter, did David getting through the PhD program at all costs serve any kind of educational objectives. Despite her frustrations with David's dissertation, Susanna was sympathetic to his situation. Given the limited scientific and language training David had received in his home country, Susanna believed that a PhD had always been an unrealistic and inappropriate target for a student allowed only three years at IFF. In her analysis, the promise that David and Peter would leave Songlin with PhDs was made only to impress international project partners and was made with little regard to the interests of the students or their research. Fulfilling unrealistic promises made by IFF's leadership was, unfortunately, typical of the kinds of headaches that junior research staff like Susanna had to endure.

Contributing to Susanna's lack of enthusiasm for the process was the fact that PhD defenses at IFF were notorious among staff for the somewhat facile questions that examiners would pose to candidates. In one case, an examiner's only contribution to proceedings was to interrogate the formatting of page numbers on the dissertation's contents page. Such instances contribute to a sense in which IFF-China's PhD defenses are more generally a meaningless formality. In some instances, however, IFF staff instead view PhD defenses as hugely valuable sites for receiving important feedback on research. Indeed, in David's case, Susanna remarked that the feedback they received during his defense turned out to be surprisingly good and would help her and her student further improve their mining restoration research before submitting it for publication. Here, Susanna rendered the interruption of the PhD defense as something more than the headache she had initially anticipated: the feedback provided at the defense became something that would shape future work in a productive manner.

This was an aspect emphasized by other colleagues at IFF. Kumar, an IFF mycologist, compared the facile questioning he saw at defenses in China to what he had seen at Chao Phraya University, a Thai institution where many of IFF's affiliated PhD students are enrolled. Among the differences Kumar noted was the superior quality of the feedback. At Chao Phraya defenses, Kumar told me, "the comments are really sharp." Where feedback at a PhD defense intervenes to affect the future direction of a research project, it epitomizes a broader notion of scientific interaction. As described in the previous chapter, IFF scientists often publish research in the hope of shaping the future direction of research conducted by colleagues in a wider scientific community. The goal that Matt and his colleagues set for their review of Asian edible mushrooms, for instance, was to "provoke activity" in Asian mycology. Like David's external examiner, Matt and his colleagues sought to make a positive intervention in the trajectory of scientific colleagues' future research. What differentiates the positive valence of these forms of feedback or intervention from, for example, the headache of the MELA framework or a visa application is how, having interrupted the trajectory of scientific work, the former might positively shape subsequent research.

Here, then, we see the potential for a PhD defense to be read alternately as a meaningless formality—as a headache—or as a valuable opportunity for scholarly feedback and conversation. Even in its more positive rendering, a defense is still a moment of interruption. Prior to David's defense, he and Susanna already had an idea for the content of journal articles that they would publish. But feedback from David's examiners caused him and Susanna to revise these plans. Rather than a mere headache—an interruption that is nothing more than an inconvenient pause—the interruption of David's defense caused a positive change in the direction of subsequent work. The difference between the valued interruptions that punctuate scientific research and the meaningless interruptions that constitute administrative headaches is usually taken for granted by IFF staff. As David's defense demonstrates, however, this ostensibly self-evident distinction is unstable: meaningless headaches can become valued opportunities for feedback and vice versa.

Another example of the contingent and unstable boundary between meaningful feedback and meaningless headache is peer-reviewed publishing. In this respect, when IFF colleagues publish in academic journals, interrupting the path of others' scientific knowledge production is seldom the only motivation. Publishing can also serve as a significant means for, among other things, establishing one's credentials for the purposes of job applications and promotions as well as for demonstrating to donors that quality research has been produced. This was something IFF managed to circumvent with MELA by ensuring that evaluation was measured by numbers of "knowledge products" rather than numbers of "publications." This deliberately ambiguous metric left open the option to report self-published or even unpublished research reports. Whereas in that instance IFF scientists were able to set themselves a relatively soft target, evaluation is more often tied to the somewhat stricter measure of *peer-reviewed* publications. In a proposal for a grant to conduct mycological research in Myanmar that would complement ongoing research in China as well as the Agroforestry for Myanmar project, IFF promised the prospective donor eight peer-reviewed publications. Promises and demands of this kind generate an imperative for what Susanna described to me as "publications for the sake of publications." "Salami slicing" is a notorious response to such demands. This is a tactic that entails parsing up data from a single project into separate papers, even though a single comprehensive paper might in other respects have made more sense. Susanna feared that, if funded, this latest mycology research proposal would leave her with no option but to indulge in some salami slicing of her own. Here Susanna identified a potential for a significant portion of publishing work related to the proposed Myanmar mycology project to become a frustrating and meaningless headache.

Even in the absence of demands for project deliverables, there is often a sense in which scientists hope that peer reviewers will offer only praise and pave the way for a manuscript to proceed promptly and unrevised to publication. No less than anthropologists, IFF staff are somewhat ambivalent about peer review. As much as they might harbor hopes for the ever so rare "accepted with minor revisions" at the first attempt, IFF scientists often nevertheless appreciate the value of peer-review comments for enhancing papers that editors determine require more significant revisions. A good reviewer might point to additional publications that the author might reference, highlight implications from research results that the author has underdeveloped or even overlooked, or indicate ways in which methods and finding could be presented with greater clarity. Conversely, there are many instances where IFF staff complain of peer reviewers who have offered unhelpful and at times somewhat venomous feedback. Nevertheless, peer reviewers can and often do positively interrupt the process of writing up research, helping scientists enhance the quality of their published work.

Of course, these interruptions are easier to take seriously and to value if the authors view the manuscript itself as a meaningful output and not simply as part of a salami-slicing ruse to satisfy a demanding supervisor or research funder. IFF colleagues gaming evaluation regimes by salami slicing provides a further example of the tendency Michael Power (1994, 16) describes wherein audit ultimately "draws organisations away from their primary purpose." The motivations that drive IFF scientists' efforts to game evaluation systems are a long way from provoking research activity on Asian mushrooms or from transforming rural landscapes. This is a phenomenon that Jiaolong raised in her blog criticizing scientists' failure to communicate with audiences beyond the sciences.[7] Jiaolong attributed the apparent lack of enthusiasm that many of her colleagues have in interesting broader audiences in their research to the necessity to "publish or perish." Rather than an aspiration to produce and share knowledge of interest and significance to other members of society, Jiaolong suggests that scientists' energies have been captured by the imperatives of professional survival. In this rendering, the work of gaming the system to satisfy contemporary modes of professional evaluation ceases to be an occasional interruption and becomes the very core of professional life. Overwhelmed by the persistent demands of managing relationships with project partners and negotiating M&E mechanisms with project funders, the headaches Matt deals with lost their character as interruptions, becoming instead the core of his work. Likewise, with publish or perish, persistent and excessive institutional demands for publications erase the prospect for scientists to do the kinds of work that Jiaolong believes they should be doing (also Mountz et al. 2015). Publication becomes a time-consuming box to tick, one that eats up time to do things that might much more effectively communicate scientific knowledge.

There are therefore two senses in which audit and the various other bureaucratic headaches endured by IFF scientists transform the interruptions of conventional scientific work. In the first sense, audit erodes the value of interruptions like a PhD defense or a journal's peer-review process to transform the trajectory of research in positive and unexpected ways. In the shadow of pressures such as those to publish at increasingly improbable rates, peer review can become a meaningless headache that, like the process for procuring official government research permissions, must be dealt with as efficiently and quickly as possible. In the second sense, the headaches that audit generates proliferate to the extent that scientific work as IFF scientists conventionally imagine it—or imagine it should become—enters a permanent state of interruption; time for the activities that Jiaolong, Matt, and their colleagues see as worthwhile evaporate from the horizon.

From Autonomy to Interruption

Building on Power's (1994, 16) argument that audit often "draws organisations away from their primary purpose," Shore and Wright suggest that there is also a sense in which the peculiar rankings and measurement systems used in universities have ended up undermining audit's own primary purpose. Shore and Wright (2015, 423) argue that despite ostensibly being introduced as a mechanism for developing trust in universities, audit has "[p]aradoxically . . . increase[ed] levels of mistrust, as trust and professional judgments came to be replaced by formal systems of auditing and inspection."[8] For Shore and Wright, trust is also at the forefront of how we should respond to audit. They ask: "How can we reclaim the professional *autonomy* and *trust* that audit practices appear to strip out of the workplace?" (Shore and Wright 2015, 422, emphasis added).

In the wake of anti-expert populism, questions of trust and of the sciences' relationship to the public are as pertinent as they have ever been (Collins, Evans, and Weinel 2017; Jasanoff and Simmet 2017; Sismondo 2017a, 2017b). But is this a challenge that we should pair with the reclamation of autonomy? Doubtless this is a connection that would appeal to colleagues at IFF. Demanding to be left to get on with one's work autonomously has intuitive appeal as a response to unending interruption. To desire autonomy, however, overlooks the arguments that STS scholars have long made not only for the empirical fact of the sciences' interdependence with the world around it but also for the embrace of the sciences' embeddedness in the world as a political and ethical virtue (Latour 2004a; Haraway 1991; Harding 2015; Liboiron 2021; Roy 2008). In fixating on autonomy, moreover, we risk forgetting the value that scientists already recognize in positive forms of interruption such as peer feedback.

Significantly, in the context of the sciences' relationships with communities outside professional science, the feedback that IFF scientists seek is not exclusively from scientific colleagues. The Agroforestry for Myanmar project, for instance, sought to engage local farmers in the design of the experimental agroforestry plots that IFF was setting up. When it launched the project, the IFF team had already produced an initial design for the configuration of trees planted in the plots, but it wanted to involve local farmers in selecting the crops. As such, the IFF project team planned for two Myanmar NGOs involved in the project to run workshops through which local farmers could select crops to be planted in the experimental plots. Various headaches that created delays in IFF's headquarters signing agreements with these project partners and transferring funds to them, however, meant that they could not set up the workshops in time for planting. More than a meaningless interruption, these bureaucratic headaches destroyed an important part of the project plan. They destroyed opportunities for local communities to interrupt IFF's research. In this instance, the IFF project team had not wanted to get on with crop selection autonomously and entirely undisturbed by external actors. What bureaucratic processes for formalizing agreements and fund transfers robbed from the project was the potential for farmers to give feedback and input into the research process.

A more successful effort by IFF to create opportunities for interruption occurred prior to an experiment that investigated the effects of various management techniques on matsutake yields. In this instance, the project team—of which I was a member—ran a brief workshop with local mushroom harvesters prior to implementing the experiment. The format of this workshop involved our team describing the proposed experiment to local harvesters whom we had recruited to implement the experiment. As well as providing an opportunity to introduce harvesters to the protocol they would be responsible for implementing, this workshop served as an opportunity to solicit feedback on our proposed research. Much as was the case with the Agroforestry for Myanmar project, our motivation for doing this was to increase the likelihood that our research findings might be of relevance and of use to communities engaged in mushroom harvesting. During the workshop, harvesters raised an objection that one of the management techniques we proposed to investigate—covering young mushrooms with a plastic dome—would increase the likelihood of mushrooms being found by poachers. The simple solution that we agreed on to address this concern was to record instances of mushroom theft as a part of the research protocol. Much as Susanna celebrated the productive feedback David received at his defense, we were extremely glad of this result. To have harvesters interrupt and, even if only in this limited manner, modify the direction of our research was something that we believed enhanced our research.

This positive valuation of farmer interruptions—in both the Agroforestry for Myanmar project and the mushroom experiment—suggests the embrace of a logic that is familiar to the outcomes thinking philosophy that Lesley introduced in her Theory of Change workshop. It also mirrors a philosophy that has a long history in participatory agricultural research (Chambers 1983; cf. Cooke and Kothari 2001). In chapter 1, I described how the outcomes thinking logic of the Lengshan Multi-Stakeholder Platform demands that scientists imagine themselves as the agents of provisionally determined transformations. At the same time, I highlighted a collaborative dimension of outcomes thinking whereby platform members from diverse walks of life are invited to shape collective research and development work. As part of the initiative to establish the Lengshan Multi-Stakeholder Platform, IFF hosted a training workshop on the theory and practice of multistakeholder innovation platforms. The workshop was delivered by two scientists, Adem and Dennis, from one of IFF's Africa-based sister organizations to an audience from a variety of agricultural research institutes and NGOs who were engaged in the Lengshan Multi-Stakeholder Platform or in similar projects elsewhere in Asia. A key theme of the workshop was Adem and Dennis's claim that agricultural research for development is most effective when farmers play a central role in setting priorities for policy and for research.

To some extent, the rationale for participation of this kind is an instrumental one: farmers are more likely to trust research findings and act on them if they feel a sense of ownership in the project. In this respect, Dennis described to me how his own research had shown that farmers would trust IFF research only if they were involved in the process from the beginning. But having explained this finding, Dennis also suggested a sense in which empowering local farmers to shape development projects is an end in and of itself. He described a multistake-holder platform he was involved in where he and his colleagues had built "so much trust that the participants would do anything he suggested" (cf. Wynne 1992). For Dennis, this was a sign of the project's failure. The project participants' excessive deference to Dennis reflected his failure to create an environment for stakeholders to shape research and development activities. Here, Dennis had not wanted the platform to simply be a vehicle for agricultural scientists to act on other stakeholders; he had also wanted it to provide opportunities for other stakeholders to act on agricultural scientists. Where project participants merely deferred to scientists—what Dennis described as an overabundance of trust—this goal was frustrated. As with the crop selection process for the Agroforestry for Myanmar project, the aspiration was for an aspect of the project to emerge as the effect of input from beyond the scientific community: for stakeholders to shape research design and scientific practice. Just as feedback from scientific col-leagues conventionally shapes the direction of research, participatory and col-laborative research models seek feedback from broader communities.

In *The Method of Hope*, Hirokazu Miyazaki (2004) argues that Suvavou people's gift-giving and government petitioning practices turn on the ability to place one's agency in abeyance and to thereby generate indeterminacy. As Mi-yazaki (2014, 84) emphasizes, "indeterminacy is not . . . a given condition, but a condition to be achieved." He suggests, moreover, that the structure of Suvavou gift exchange and government petitioning can provide a model for our own eth-nographic practice. More recently, he has contrasted this implicit Theory of Change for ethnography with Casper Bruun Jensen's (2014) call for anthropo-logical "insistence": "Anthropological commitment to analytical openness re-sides in the cultivation of an outward orientation toward, and a willingness to receive and respond to, others—whether they are fellow anthropologists, research subjects, or collaborators—rather than in the cultivation of an internal strength to go on as anthropologists" (Miyazaki 2014, 526).

Dennis's Theory of Change implies an aspiration to a similar mode of open-ness. This is an openness that entails moments of willful abeyance that antici-pate the resumption of a research project shaped in unpredictable ways by those who have interrupted it. According to Dennis's implicit Theory of Change, as important as a scientists' ability to shape the future is their ability to cede some control over a project to rural communities. In contrast to what he inadvertently achieved—the establishment of scientists' autonomous authority—Dennis's goal is a relationality that would allow scientists to place their agency in momentary abeyance and with that generate the indeterminacy that is fundamental to collaborative research.

Trust in Science

IFF colleagues experience bureaucratic and administrative work from visa applications and acquiring permission for field research to donor monitoring frameworks and project reports as headaches—unwelcome interruptions that will ultimately make no difference to the meaningful work that they punctuate. Efforts to overcome headaches can be animated by quite different, and at times quite fragile, horizons—from the promise that bureaucratic work will subside once one's academic career progresses to fears for a future of endless headaches. In some cases, IFF colleagues might experience a single process—a PhD defense or peer-reviewed publication, for example—as a headache in one instance and as a welcome opportunity for feedback in another.

Much like IFF colleagues, many anthropologists view the headaches that emanate from funding bureaucracies as radically distinct from the valued feedback that we receive from academic peers and from fieldwork interlocutors. Understood as alternative forms of interruption, however, the difference is not so stark. The distinction lies merely in the kinds of future horizon for one's work that these interruptions anticipate—a future delayed but not transformed or a future that might transform in ways one cannot fully predict. Understanding science and bureaucracy in this way might challenge us as scientists to reflect on the temporal frameworks—the implicit Theories of Change—within which we imagine our engagements with audit cultures and with academic bureaucracies. Doing so, we might find ourselves surprised in the manner that Susanna was by David's defense: we might find opportunities for transforming the pathways of our research where we would otherwise only anticipate headaches. Writing grant reports, for example, we might not merely imagine these activities as meaningless headaches but also open ourselves up to the question of how speaking to the audiences of these documents might transform the future direction of our work.

More importantly, shifting focus from the representational paucity of academic bureaucracy to the quality of interruptions they facilitate also allows for its own critique. From the perspective implied by focus on interruption, the problem with funding reports might not so much be that they provide a poor means of representing the work that we have done or will do. The problem, rather, is that as hard as we might try to imagine ourselves in conversation with the audiences of these documents, these audiences seldom speak back in productive ways.

Far from the unpredictable interruptions that IFF scientists value from each other and from local project participants, the interruptions generated by grant and fellowship progress reports (and in some cases by the applications we write for them) are often one of only two kinds: they either end a project or allow it to proceed. Under such circumstances, it is little wonder that scientists only see the interruptions of donor bureaucracies as headaches. In many cases, they do not offer much else. A polite, boilerplate email telling us only that we have or have not won a fellowship or thanking us for completing a project report is an unlikely basis for unexpected transformations in a project's future.[9] As well as interrogating the inadequacy of the interruptions that audit regimes generate, we can also see the capacity for them to destroy opportunities for fruitful interruption. In this regard, the cancellation of farmer involvement in crop selection for the Agroforestry for Myanmar project demonstrates how bureaucratic and institutional hurdles can frustrate modes of interruption that are vital to scientific practice.

Interrupting and being interrupted by our interlocutors may of course often entail us making some version of our perspectives visible to one another. But to focus on the opportunities for interruption that audit and research bureaucracies can and should generate is a different kind of critical exercise to that of explicating what is ultimately made (in)visible by an evaluation regime. What is at stake here is not only that scientists or development professionals have lost control of the terms of their evaluation or that the work of constructing reality has been co-opted by powerful others. Rather, we should see the problem of audit and evaluation frameworks in terms of the interruptions that they generate and the failure of these interruptions to facilitate the potentially enriching indeterminacy of momentary submission to the agency of others.

The chapters that follow continue to explore the diverse forms of interruptions that are fundamental to research and development practice at IFF. In doing so, I hope both to show what some of the contours of a better-interrupted science might look like and to further flesh out what we might gain by thinking of scientific practice in terms of interruptions. Importantly, re-posing our response to audit in such terms suggests that public trust in the sciences cannot be tied to scientific autonomy. When Dennis laments project participants who trust too much, he describes a relationship of unquestioning confidence in and allegiance to his recommendations. He had generated a relationship wherein project participants saw no need to interrupt him. This is a form of trust in scientific knowledge that leaves Dennis's and his expertise all too autonomous. As he points out, this is not the kind of trust scientists should seek. The chapters that follow ask what alternative form trust in science could take.

REFUSING TRUST, EVADING VULNERABILITY

IFF's China office has a dual legal identity: it is the regional office of an international organization named IFF, and it is a research center within a Chinese public research institute named the Songlin Academy of Plants Sciences (SAPS). The former of these identities compels scientists to mold their work to the ever-changing policies of IFF's East African headquarters—including agendas for impact and value for money described elsewhere in this book. The latter identity requires these same scientists to work within the many regulations that govern public bodies in China. IFF staff must, for instance, comply with spending limits and audit requirements introduced as part of a nationwide anticorruption campaign. This campaign has included a series of high-profile cases against top-level government officials as well as numerous prosecutions of lower-level officials. Of greater significance to the day-to-day work of IFF staff, however, are the litany of measures introduced to curb the extravagance of civil servants. The consequences of these regulations are clearly visible to IFF staff. "Before," Mubai told me, "whenever we met with local government officials, there would be a big banquet and lots of liquor. Things are much more modest now." For the many IFF staff who collaborate with local government on research and development projects, this is quite a welcome change. Being cajoled into excessive drinking was not something many IFF colleagues looked forward to. Mubai, for instance, used to dread visiting certain field sites because he knew that going there meant invitations to liquor-fueled dinners. Beyond instigating frugal dining and more moderate drinking, anticorruption measures have brought several less welcome consequences. There are now, for example, strict spending limits—combined with stringent accounting procedures—governing the transport, accommodation, and food expenses that staff can claim for field research and academic conferences. Seemingly intended to prevent indulgences like the expensive liquor and culinary delicacies for which state officials are notorious, the amounts allowable are so small that they would make many of IFF's research activities impossible.

Often, though not always, IFF staff are able to circumvent the difficulties created by such expenditure limits. In some instances, the problem is simply the unavailability of official receipts known as *fapiao*. These receipts are required for reimbursement on expenses such as taxi fares or the purchase of research materials. Only registered businesses can issue these receipts, but even businesses that are registered do not always carry receipts because businesses must pay tax on transactions for which they are issued. In one instance, Mubai and I were traveling in a taxi across numerous remote research sites in Lengshan Prefecture. It emerged during our travels that the driver did not have any receipts to give us. As an alternative, Mubai suggested that the driver sign a form stating that he had been working for IFF as a research assistant. Reporting expenses in this way would allow Mubai to report up to CN¥200 (US$30) per employee per day. In this instance, CN¥200 per day would not quite cover our taxi fare, but Mubai had a remedy to this problem. He asked the driver to provide a member of his family as an additional fictional research assistant. The difficulty, however, was to persuade the driver to share his and a family member's government ID numbers and to thus entangle himself in Mubai's fiction. In this instance, the taxi driver refused to share the ID numbers, preferring instead to take us on a two-hour round trip to procure the necessary receipts.

In this chapter, I analyze bureaucratic workarounds of the kind Mubai attempted here as well as some of the reasons why a taxi driver might be reluctant to be party to them. I take my lead in this analysis from a colleague named Tao's frustration with what she calls bureaucrats' "refusal to trust."

Refusing to Trust

As a research assistant in Matt's soil biology group, Mubai was responsible for navigating all manner of bureaucratic headaches—most frequently doing so on behalf of foreign researchers such as myself. The greatest bureaucratic burden, however, falls on a corporate services team led by Tao. Over a lunchtime coffee, I asked Tao to talk me through some of her experiences with SAPS's anticorruption measures and financial reporting requirements. "I hate having to deal with these stupid rules," she told me. With a sigh that expressed months of frustration dealing with impossible bureaucratic procedures, Tao described the experience of a friend she knew in another of SAPS's research units. This friend had been tasked with overseeing an international academic conference in Guangzhou. SAPS's anticorruption regulations dictated that in hosting the conference, the organizers could not spend more than CN¥350 (US$53) per conference attendee per day. In a city as expensive as Guangzhou, Tao explained, this amount is laughably small. There was simply no way the research unit in question could cover expenses for facilities and food with such a tiny budget. Tao lamented the impossibility of asking SAPS administrators to apply the spending limits flexibly. Despite the manifest impossibility of anyone organizing a conference within the mandated CN¥350 limit, there was no prospect, Tao explained, of negotiating an exception to the rule. Tao's friend had a simple solution: He would fabricate fictional conference attendees. This meant that although the conference's expenditure was greater than the amount allowable, the addition of fictional attendees to the financial report created the appearance of having satisfied per person expenditure limits.

The reason that workarounds like fictional conference attendees are necessary, Tao explained to me as she sipped her coffee, is that rigid and inflexible application of the rules allows bureaucrats to sustain a detachment from their clients' rule bending. Tao described this as a "refusal to trust." For Tao, this refusal to trust is intertwined with fears that even the most minor rule breaking or wrongdoing could serve as a pretext to rebuke. Like the Pakistani bureaucrats described by Matthew Hull (2012) and the Mexican forestry officials described by Andrew Mathews (2011), Chinese bureaucrats are acutely conscious of their vulnerability. Fear of the consequences of being caught up in irregular activities, Tao explained, drives administrators' and bureaucrats' refusals to trust their clients. In this explanation of "refusing to trust," Tao draws a connection between trust and vulnerability: to evade trust is to evade vulnerability. Trust, in Tao's conceptualization, would imply a willingness to make oneself vulnerable—to expose oneself to the possibility of harm—for the sake of another. For SAPS administrators to have trusted the Guangzhou conference organizers and allowed an exception to the ordinary conference spending constraints would have meant those SAPS administrators sharing in vulnerability to the scrutiny of any official who might interrogate the legitimacy of that conference's budget. What rigid rule following allowed, by contrast, was for bureaucrats to evade the vulnerability that trusting Tao's friend would have entailed.

In social scientific scholarship, trust is often used interchangeably with or as a synonym for confidence or belief (for example, Collins and Evans 2002; Ialenti 2020; Rothstein and Stolle 2008; Yang and Tang 2010). But understood as a willingness to make oneself vulnerable to or on behalf of another, trust is something distinct from confidence or belief in the propensity of another to tell the truth or to provide reliable knowledge. The concept of trust underlying Tao's analysis is perhaps closer to theories that connect trust to risk (Ingold 2000; Luhmann 2000; Rabinow 2011) or to vulnerability (Chua, Morris, and Ingram 2009; Rousseau et al. 1998). Heeding Alberto Corsín Jiménez's (2005, 64; 2011, 178) call for us to "resist the temptations of a sociology of trust," however, an etic definition of trust may not be the most productive starting point. Indeed, rather than intervene in or contribute to efforts to build a general theory of trust, my goal is to elucidate the specific concept of trust that Tao articulated to me and to realize the potential of that concept to illuminate the worlds of bureaucrats and their clients. It is with that goal in mind that I turn to the anthropology of gifts—a literature that, like Tao's analysis, reminds us that risk, hazard, and vulnerability are inherent to relationships (Keane 1997; Mauss 2016; Nadasdy 2007; Raheja 1988; Weiner 1976).

It is the hazards and entanglements of social relationships, according to James Laidlaw (2000, 631), that give the idea of "overcom[ing] the impossibility of a free gift" its allure. To receive (or to give) a free gift would be to achieve a transaction that incurs no debts and creates no connections. It would be to transact in a manner that generates none of the danger attendant to ordinary social interaction. Laidlaw identifies an institutionalization of this apparent impossibility in the process by which Jain renouncers receive alms. The spiritual purity that Jain renouncers pursue forbids them from manual work and from preparing their own food. It also forbids them from partaking in reciprocal exchange. Jain renouncers are as such reliant on alms. The way in which renouncers collect alms takes a very particular form that is known as "grazing": a metaphorical reference to the way livestock wander feeding on no particular patch of grass. When collecting alms, Jain renouncers wander aimlessly with their alms bowls, receiving food from local households but never accepting an invitation or offering a thank-you. This protocol erases the potential appearance of exchange or reciprocity, allowing Jain renouncers to sustain the alms they receive as an impersonal free gift. In doing so, Jain renouncers ensure that they are not entangled in any kind of relationship with the people from whom they receive food. These are transactions that, as Laidlaw (2000, 617) puts it, "mak[e] no friends."

The detachment and autonomy that Jain renouncers strive to sustain through grazing are tied to their pursuit of spiritual purity. For the administrators that Tao admonishes, the pursuit is somewhat less spiritual, but what is at stake is similarly the maintenance of detachment from social relations and the dangers that they entail. Like Jain renouncers, SAPS administrators employ decidedly impersonal modes of transaction as means to evade the entanglements, dangers, and vulnerabilities inherent in interpersonal connection. Flexibly applying

SAPS's impossible expenditure rules would entail a personalized transaction—one in which a bureaucrat's discretion would tie them to the rule bending of their clients. Rigid proceduralism, by contrast, offers the same bureaucrat the opportunity to avoid (or at least attempt to avoid) any personal connection to their clients and with that the possibility of a share in responsibility for their actions. Refusing trust is, in short, all about shoring up personal detachment and autonomy.

Anthropologists have long marveled at the ability of "the Chinese . . . [to] tur[n] strangers into connections" and at the apparent imperative for people in China to expand their network of social connections (*guanxi*) at every opportunity (Hertz 1998, 25; also Fei 1992; Hathaway 2013; Osburg 2013; Yang 1994; cf. Chun 2018). IFF staff's frustrations with SAPS administrators and government officials, however, point to a situation where people are no less anxious to keep their interlocutors at arm's length. The desire that motivates fastidious rule following and excessive demands for documentation is, to borrow from Laidlaw, for a transaction that makes no social connections.

Getting Sent in Circles in Damei

Bureaucrats' refusal to trust—their desire for a transaction that makes no connections—is evident in the struggles that IFF staff endure to gain access to field research sites. This is perhaps especially so when it comes to IFF's many foreign researchers, not least me.

One of the main tasks Matt assigned to me as an IFF research assistant was to conduct social science research on wild mushroom harvesting. At the heart of this project was a household survey I conducted across various sites in southwest China. As with my trip to Lengshan—and as is the case for most international scientists at IFF—I was fortunate to have the support of a local colleague accompanying me on the first of the field trips I made to implement this survey. This time my field guide was Kong Shifu. Though his official position was the office's driver, for myself and many other international scientists at IFF, Kong acted as a de facto and indispensable expert in navigating local bureaucracy. As importantly, his sense of humor and his ability to find fun in the most unlikely of situations supplied much-needed respite during often-exhausting travels to field sites across rural southwest China and Southeast Asia.

After a long day's driving to our first field site—a Tibetan village named Meirong—Kong suggested to me that it would be easier for me to implement our household survey if we had the cooperation of local government leaders. As such, rather than go directly to Meirong, Kong first took us to the offices of the Damei

County Government, where he sought out the county director. Handing over an official letter of introduction from SAPS, Kong explained our intention to conduct research on mushroom harvesting. He asked the county governor if he might assist us with an introduction to the head of Meirong Village. The county director politely responded that he would be happy to help but explained that because our party included a foreigner, we must first present ourselves to the Damei County Foreign Affairs Office. The Foreign Affairs Office, the county director explained, would be able to issue a stamped document approving our request to conduct research in Damei. Once we had this document, the director continued, we should bring it back to him, and he would call ahead to Meirong to introduce us and ask the village leader to accommodate us.

When we arrived at the Foreign Affairs Office, however, two junior officers told us that our SAPS introduction letter was insufficient. They requested a more detailed letter from SAPS laying out, among other things, the specifics of our research plan in Damei and indicating which SAPS professor was ultimately responsible for the project. Kong promptly made a phone call back to IFF's office in Songlin, where IFF colleagues explained that they would chase SAPS administrators for the necessary documents. In addition to requesting further documentation from SAPS, the Foreign Affairs Office informed us that although they were the responsible office for foreign visitors, they did not have the authority to approve a project related to mushroom harvesting. This, they suggested, would be a matter for the County Forestry Bureau. In addition to the supplementary documentation from SAPS, we should, the officers told us, ask the Damei County Forestry Bureau for a stamped letter approving our survey in Meirong and then bring that letter back to the Foreign Affairs Office. By this point, Kong and I were not surprised to find help hard to find at the Forestry Bureau. A junior official turned us away, telling us that everyone was out at lunch. While we were waiting for the Forestry Bureau officials to return from their lunch, we received a call from Yanli, a member of IFF's corporate services team who had been working to secure the additional documentation that we needed from SAPS. Yanli explained that the documents we were after would require the signature of SAPS's director but that he was not on campus that day. Yanli therefore suggested that we simply try our luck going straight to Meirong. If we were turned away, she told us, we should simply move onto our next research site. That site was a half-day's drive back in the direction we had come from but was in a non-Tibetan area where we expected fewer obstacles to our research.

On our way to Meirong after lunch, Kong suggested we drop in at the Forestry Bureau just in case they might help us out. Kong instructed me to stay in the car, lest the sight of a foreign researcher make the officials there nervous. To our surprise, the bureau director was quite happy to help and gave Kong his phone number, telling him that he could call him if we encountered any difficulties. When we arrived at the village, however, we were disappointed to find that this phone number did not satisfy the village leader. He told us that he did not know anything about the Damei Forestry Bureau and that he needed approval from his direct senior in the township government (the level below county). The officials at the township government were similarly unimpressed by the phone call that we offered them or by our existing SAPS introduction letter. They told us that before returning to their township, we must acquire written permission from the office where we had begun several hours earlier: the Damei County Government.

Having wasted the day being sent from office to office in a futile effort to acquire official approval for our household survey, Kong was losing patience. With over a decade's experience as IFF's driver, Kong was no stranger to the headaches we faced in Damei. All the running in circles was nevertheless beginning to test his jovial demeanor. Not wanting to waste any more time chasing our tails, Kong decided to make a call to Professor Lin—a senior member of staff at IFF—wondering if she might have any ideas. As it happened, a university classmate of Lin's was now a senior civil servant at the local prefectural (the level above county) forestry bureau. Lin made a call to her classmate to explain the predicament. When we returned to Meirong the next morning, the village leader greeted us warmly and assigned one of his officers to assist us in the household survey. Having wasted a day being bounced from one office to another, a single phone call erased all our difficulties. The simplicity of the resolution reflects a familiar wisdom: that one often "gets one's way not by observing formal and bureaucratic regulations or by going through the proper channels, but by creatively seeking out unofficial routes, detours, and shortcuts" (Yang 1994, 130). But what of the documents that each of these offices demanded? And the efforts of our colleagues back in Songlin to procure these documents?

Documents That Open Doors

Echoing Tao's lament that bureaucrats refuse to trust, Kong and his colleagues at IFF put our tail chasing in Damei down to officials wanting to avoid responsibility for authorizing the presence of a foreign researcher in Meirong. This evasion was tied, as IFF colleagues see it, to bureaucrats' fear that a foreign researcher might make trouble—a perception that is especially heightened in the notoriously sensitive context of Tibetan regions like Damei. Our being bounced from office to office was in many respects a familiar example of bureaucratic "buck-passing" (Herzfeld 1992, 4). The paperwork fixation of the bureaucrats we encountered also points to the potential that Matthew Hull (2012; also He 2012; Mathur 2016) highlights for documents and files to diffuse agency and to guard against individuated responsibility. In his ethnography of Pakistani bureaucracy, Hull argues that while external observers lament the authorless nature of the bureaucratic document, for insiders, authorship is all too precisely specified. Indeed, guarding against individual authorship of an official decision is a central goal of Pakistani officials' document-mediated bureaucratic practices. Hull, much like Tao, connects this to a sense that bureaucrats are under a constant threat of scrutiny. Anxiety about this presumptive threat is heightened by the tendency of successive national governments to purge bureaucrats, by threats of demotion and dismissal from the senior ranks of the local civil service, and by the ongoing activities of state audit and investigative bodies (Hull 2012, 127–28). This echoes Andrew Mathews's (2011, 160) ethnography of Mexican forestry officials, where he describes bureaucrats animated by a "profound sense of frustration, doubt, and personal vulnerability." In such a context, Hull argues, the central task of bureaucratic activities becomes the construction of a diffuse and anonymous collective agency. As Hull (2012, 115) puts it, "the intensification of file-mediated decision making undermines the ability of superiors to isolate individual functionaries and hold them responsible for particular actions."

The personal phone call that brought an end to Kong's and my bureaucratic travails in Damei is the familiar informal alternative to the intricate paperwork that Hull describes. Rather than documents that diffuse agency, we found in Lin's classmate at the Prefectural Forestry Bureau an official willing to assume individuated responsibility for authorizing our research. The background to this was, however, a group of bureaucrats pursuing a documentary diffusion of agency and responsibility in a very similar manner to the Pakistani bureaucrats described by Hull. At each juncture in Kong's efforts to acquire research approval, the officials we met sought to diffuse the potential for individuated responsibility by collectivizing the process of approving our research. This process was, as in Hull's ethnography, to be mediated by a proliferation of paperwork. A document from the Foreign Affairs Office would have allowed the County Government officials to diffuse their potential responsibility for the foreign researcher, just as a document from the County Forestry Bureau would have done for the Foreign Affairs Office officials. Equally important in this process was the demand for robust paperwork from SAPS administrators who Damei officials would thereby ensure shared in responsibility for authorizing our research. This documentation would, in other words, facilitate a transaction that makes no individuated connections.

We should not ignore this documentary work simply because Lin's personal connection later made the documentary procedures appear irrelevant. Indeed, to consider only the back-door solution to our problem would be to ignore the labors of colleagues back in Songlin who pursued the documentation that the Damei County Government and the Damei Foreign Affairs Office had requested. Though a significant headache, these documents were not dead ends. Rather than a lack of confidence that we could meet these bureaucrats' documentary demands, our main concern was with the amount of time it would take to procure these documents. We were in Damei with a team of four survey enumerators with nearly three hundred household survey questionnaires ahead of us. Each day we spent in Damei chasing documents meant not only wasted time but also wasted wages and wasted hotel bills. The back door provided by Lin's connection spared her IFF colleagues the time and the labor of pursuing troublesome paperwork, but personal connections were by no means the only tactic that IFF staff could pursue in such circumstances.

In many other instances, IFF staff do satisfy the demands and anxieties of Chinese bureaucrats by ensuring that IFF adheres to—or at least appears to adhere to—clearly set out rules and document-mediated procedures. Though a letter of introduction from SAPS did not satisfy officials in the Tibetan Damei County, in most parts of rural China, a similar document is all that is needed to convince local officials to assist the letter holder. Without one, however, a researcher is likely to experience significant difficulties. This was the case for Frankie, a North American PhD candidate who wanted to conduct research with mushroom harvesters in rural Luobo Prefecture. Frankie's data collection was initially hampered by local officials telling her she had no business in the area and spreading rumors that she was an ill-intentioned journalist. Frankie was affiliated with IFF as a visiting researcher and worked closely with IFF colleagues on certain aspects of her research. Having had a miserable time being repeatedly rebuffed by officials and locals in Luobo, Frankie returned to Songlin to ask her IFF colleagues for help. An IFF colleague suggested that the next time Frankie visited Luobo she should take a SAPS letter of introduction. This letter, they hoped, might help Frankie establish her identity as a legitimate researcher.

Mubai was sympathetic to Frankie's predicament and took it on himself to help get Frankie the letter she needed. He had, moreover, plenty of experience helping members of Matt's team find and access field sites. This included routinely procuring stamped SAPS letters of introduction. Just as he had done for countless members of Matt's team, Mubai wrote up a short summary of Frankie's research plans in Luobo. He then took this to the SAPS office that oversees IFF and asked for a stamped letter of introduction that Frankie could use to request support from officials in Luobo. The SAPS administrators Mubai spoke with, however, refused to oblige his request. Just like the officials in Luobo, it seemed SAPS administrators also wanted to keep themselves at arm's length from a foreign researcher.

Though Frankie was indeed collaborating with IFF and was in China on a research visa that SAPS had sponsored, the SAPS administrator explained to Mubai that they would only issue letters of introduction to researchers on the IFF payroll or registered for a SAPS degree program. Frankie satisfied neither of these conditions. There was, however, an easy workaround to this blockage. Mubai recruited Changkong, an IFF student registered in SAPS's own doctoral program, and had a letter of introduction made up in his name instead of Frankie's. Once Mubai had obtained this already-stamped letter, he wrote in Frankie's name alongside Changkong's. For good measure, Changkong traveled back to Luobo with Frankie to help her negotiate with local officials. Staying only a day, Changkong then left Frankie in Luobo with the SAPS letter attesting that she was conducting her research in collaboration with a prestigious public research institute. Armed with paperwork showing that SAPS was responsible for Frankie's activities, her difficulties in Luobo diminished considerably. In contrast to my experience in Damei, here it was a formal document, rather than an informal social connection, which enabled Frankie to secure the research permission she required.

Fear of Misfire

Workarounds like names added to already-stamped letters or fictional conference attendees are not especially sophisticated, nor are they unique to SAPS. Indeed, they bear similarity to the frequent "fraud, trickery, and formalism" that Zhao Shukai (2007, 65) describes in the audit systems of township government and that Andrew Kipnis (2008) observes in the audits of Chinese schools. They are perhaps familiar even to the experiences of anthropologists in our own academic institutions. Such is the prevalence of this trickery that no one doubts that SAPS administrators—and their equivalents at other public bodies—are aware of its ubiquity. IFF colleagues describe the way administrators nevertheless allow themselves to be deceived as "turning a blind eye" (*zheng yi zhi yan bi yi zhi yan*). It is, however, only under carefully managed and quite particular circumstances that SAPS administrators will oblige with a blind eye.

In one instance, Isaac, an IFF student conducting doctoral research on pollinators in coffee plantations, returned from a field research trip for which he had driven in his own car. When his colleagues at IFF submitted his gasoline receipts to SAPS, Isaac's application for reimbursement was rejected. Though his expense report was entirely accurate and there was no apparent breach of SAPS expenditure rules, SAPS administrators were suspicious of Isaac's claim to have used his own vehicle on a research trip. Puzzled by how an administrative office that routinely turns a blind eye to tactics such as fabricated conference attendees could be pushing back at what seemed such a straightforwardly honest reimbursement application, I raised the case over lunch with Mubai. To Mubai, Isaac's situation was not especially surprising. He explained that in China it is very rare for people to make a business trip in their own car, so the situation was probably unusual to the SAPS administrators concerned. These administrators, Mubai reasoned, were likely worried that this unusualness could raise suspicions with their own supervisors. Among hundreds of reimbursement claims documented via taxi receipts and gasoline for SAPS-owned vehicles, Isaac's reimbursement might stand out as an obvious candidate for closer scrutiny during any future audit. Despite being within the rules, the unusualness of gasoline receipts as a method for documenting expenses was enough to make SAPS administrators anxious.

Webb Keane (1994, 1997) has described the careful and potentially hazardous work that Anakalang exchange participants perform to demarcate alternative kinds of transaction and to thereby control the future consequences and entailments that a given transaction generates. Keane (1994, 607) gives the example of the difference between the ritual exchange that accompanies a wedding—an exchange that assures future relationships between "wife-givers" and "wife-receivers"—and "a low-status market transaction, an isolated encounter lacking spiritual or social consequences, and underwriting no future memories and obligations." Here, exchange depends on a strict adherence to the appropriate ritual protocol. If either party "bungles" the ritual speech, then the goods exchanged will lose their ability to bind the parties and to ensure their mutual dependence into the future. Keane's analysis echoes Laidlaw's concern with the future entailments of gift exchange and the vital demarcation of transactions that do and do not create connections. What Keane's analysis brings to the foreground, moreover, is the contingency of any given transaction: the possibility that a transaction might not be achieved in the form the participants to it desire. And, as importantly, Keane reveals how this contingency looms large over the participants in a transaction: the potential for "misfire" animates the careful and at times anxiety-laden performance of a transaction.

It is anxiety about misfire that Mubai attributes to SAPS administrators. Indeed, given Mubai's explanation of the grounds for SAPS administrators' suspicions, their blind eye to fabricated conference officials on the one hand and their refusal of an entirely honest claim for gasoline on the other are not as incongruous as they had first seemed. What really mattered from the point of view of these administrators was the likelihood they perceived that the reimbursement might be called into question later on: that an expense report tied to gasoline receipts might misfire, exposing not only Isaac but also the administrators to unwanted scrutiny.

In this respect, the refusal to trust that Tao highlights and that Isaac's experience exemplifies suggests a slightly different relationship between trust and audit to the one commonly foregrounded in critical and anthropological scholarship. Jillian Cavanaugh (2016), for instance, shows how audit documents serve as a form of knowledge production that allows food production companies to evaluate labor processes. She observes, however, that food production workers experience the constant evaluation of new audit regimes as an insulting display of their employer's lack of trust. Here Cavanaugh echoes an argument made by Michael Power (1997) and reiterated by several anthropologists: that "[a]ccountability mechanisms often replace relations of trust in audit culture" (Cavanaugh 2016, 696 citing; Shore 2008; also Shore and Wright 2015; Strathern 2000; Stein 2018; Tracy 2016; cf. Galvin 2018). Some put this point more strongly: transparency and audit not only replace trust, but they also "exacerbate [an existing] lack of trust" (Kipnis 2008, 281) and are "generative of mistrust" (Webb 2019, 715). On the face of it, such conclusions resonate with Tao's contention that a fastidious trail of documents facilitates SAPS administrators' refusal to trust. In analyses of audit, however, anthropologists ordinarily tie trust and mistrust to the production of information and knowledge. Specifically, the "lessening of trust" coincides with the proliferation of information (Tracy 2016, 43). Audit is insulting because it demands one provide documentary evidence for what was once taken on trust. Food workers and academics once trusted to get on with their work take offense when suddenly, they must fill out forms to prove they are doing their jobs properly (Cavanaugh 2016; Shore and Wright 2015). Isaac was perhaps insulted by what he took to be people questioning the truth of his claim to have used a personal vehicle for work. In Mubai's analysis, however, SAPS administrators overseeing Isaac's reimbursement had little interest in information that would allow them to evaluate Isaac's activities or the truth of what he told them. Understood in relation to Tao's conception of trust evasion, the primary issue is not substantial evaluation of the information contained in Isaac's reimbursement petition or of his claim to have acted appropriately.

Particularly illuminating in this regard is the advice that IFF's corporate services team offered Isaac for dealing with SAPS administrators' obstructiveness. All Isaac needed to do, his colleagues assured him, was to get hold of taxi receipts equaling the amount he had spent on gasoline and to resubmit his reimbursement claim as if he had used taxis rather than his own car. Whether SAPS administrators believed Isaac's story was beside the point. The solution to the administrators' refusal to trust was not to demonstrate his sincerity or to prove to SAPS administrators that they could have confidence in the veracity of the information contained in his paperwork. The challenge for Isaac was instead to provide a set of documents that would satisfy SAPS administrators that the reimbursement was not likely to misfire—and that, in the event of misfire, they would be sufficiently distanced from the consequences. Even though submitting taxi receipts would involve a fabrication of Isaac's activities and expenditures, this was the pathway that Isaac's colleagues believed would best assuage SAPS administrators' anxiety. Sure enough, when Isaac resubmitted his expense report with borrowed taxi receipts, he received the reimbursement he had been after. The taxi receipts allowed SAPS administrators to displace the need for "trust," not in the sense that they provided reliable or transparent knowledge of Isaac's activities but in the sense that the borrowed receipts allowed administrators to minimize their personal vulnerability to potential misfire.

Isaac's submission of fictional taxi receipts is nevertheless not merely the belated success of his initial effort to submit his actual receipts: that Isaac was forced to fabricate his expense report is consequential. For one, even if sincerity is not a core concern for SAPS administrators, it is a preferable characteristic for most IFF staff. Isaac was in no doubt that he would sooner not sign official forms stating he had taken a taxi that he did not take. Likewise, Tao lamented the situation of her friend forced to fabricate fictional conference attendees to run a Guangzhou conference. The bureaucratic system that makes such work-arounds a necessity, Tao told me, "turns innocent girls into prostitutes" (*bi liang wei chang*). Getting one's hand dirty fabricating fictions is not something IFF staff enjoy.

A consequence of authorities systematically turning a blind eye is, furthermore, the small but ever-present possibility that the irregularities to which a blind eye was once turned could be deployed as a premise for future punitive action. Mubai likened this to a fish tank you keep filling with bad fish so that the day you want to catch one it is very easy to do so. As an illustrative example, Mubai described how almost none of the electric bikes on the streets of Songlin are legally registered and they are therefore being ridden illegally. This is a consequence of the city administration refusing to register new e-bikes but nevertheless turning a blind eye to the growing presence of new, and therefore necessarily unregistered, e-bikes on the roads. Mubai suggested that if it so wished, the city government could at any point cease to turn a blind eye and have grounds to punish any of the million or so e-bike riders in Songlin. Thus, while the willingness of public bodies or government offices to turn a blind eye often allows people to evade certain rules and regulations, exploiting a blind eye can also generate a sense of vulnerability. Indeed, this is why the taxi driver Mubai and I had employed in Lengshan was so reluctant to participate in Mubai's fiction that the driver and his family had provided research assistance. Workarounds such as fictional research assistants, fictional taxi receipts, or fictional conference attendees achieve their immediate goal—expenses are reimbursed. But, as anthropologists have long understood, a transaction must be understood for much more than its effects in the immediate moment (Keane 1997; Laidlaw 2000; Venkatesan 2011; Watanabe 2015). The immediate effect of reimbursement based on actual gasoline receipts might be the same compared to reimbursement based on fabricated taxi receipts—the claimant receives a certain amount of cash—but the latter transaction generates enduring vulnerabilities for the claimant that the former would not. The cost of bureaucrats seeking to minimize their vulnerability is often to foist the vulnerability attendant to routine rule breaking on their clients. The dangers of misfire are, in this respect, asymmetrical for bureaucrats compared to their clients.

From Confidence to Vulnerability

In the process of organizing a conference, Tao imagines two pathways that would have allowed her friend to spend more than the impossibly restrictive SAPS expenditure rules permit. First, there was what Tao's friend was forced to do: to find a documentary workaround that could fit her reimbursement within SAPS administrators' proceduralist demands. Second, there was what Tao laments that her friend could not do: ask SAPS administrators to be flexible in their application of the rules. In both cases, the immediate outcome would be identical: her friend would be able to access the necessary funds to run a conference. When SAPS administrators insist on the former, however, their concern is not so much with the immediate outcome of a transaction—whether a conference can go ahead—but with managing the entanglements and vulnerabilities that transactions project into the future. This document-mediated, proceduralist relationship typifies many of Tao and her colleagues' interactions with officials within and beyond SAPS. These are relationships characterized by the refusal of bureaucrats to trust—by their determination to evade vulnerability. Rather than apply rules and regulations flexibly, bureaucrats prefer to turn a blind eye to the pervasive use of documentary fixes such as the reporting of fictional conference attendees. For bureaucratic clients like Tao, this highly proceduralist relationality generates vulnerability that extends beyond the moment of transacting expense reports and official documents. But for the bureaucrats themselves, formal transactions are carefully controlled to minimize and to diffuse their own entanglements in responsibility and vulnerability. Insofar as bureaucratic relationships generate effects into the future, these are often effects borne by the client and of little consequence to the bureaucrat, who is careful to shore up the autonomy and detachment of their own future.

As well as helping us understand bureaucratic travails in contemporary China, Tao's analysis might provide conceptual tools for thinking about a wider field of relationships at IFF and in the sciences more broadly. In the previous chapter, I described how trust became a concern for Dennis because of its overabundance: Dennis was concerned that farmers place too much trust in scientists such as those at IFF. In this chapter, I have described how trust becomes a concern because of its extreme absence: Tao's lament is that bureaucrats refuse trust. Dennis and Tao are, of course, using the term "trust" to evoke very different concepts. Dennis points to trust in the sense of confidence or belief in a body of knowledge or expertise. This is a form of trust that could lead one to blindly abide by the lessons and imperatives that a body of knowledge implies. For rural communities to have this kind of trust in the expertise of IFF's scientists might, as in Dennis's example, result in these communities unquestioningly following the advice of IFF scientists. As Dennis highlighted, this is hardly a desirable model for science–society relations. This is not the kind of trust that scientists should seek to cultivate. In this chapter, I have shown how Tao's quite different concept of trust—one tied to vulnerability—provides a critical framework for making sense of the challenges of navigating bureaucracy. But what if we also approach the problem of trust in the sciences through the lens of Tao's concept of trust?

In the chapters that follow, I will extend Tao's concept of trust as vulnerability to analyze the relationships that IFF staff cultivate with collaborators in Chinese state institutions as well as with rural communities and international funding organizations. This will mean focusing not only on the equation that Tao makes between evading trust and evading vulnerability but also on the inverse: the equation of building trust with embracing vulnerability. In extending this conceptual framing beyond the context of bureaucracy, I want to explore the potential for willful embrace of vulnerability as a model of trust for scientific practice.

HUMANIZING BUREAUCRATS

In the previous chapter, I explored IFF colleagues' critiques of bureaucracy and government officials. In addition to giving rise to a critique of bureaucrats, drawn-out administrative headaches on occasion also lead IFF staff to turn a critical gaze toward their own colleagues. What is more, the reality of the vulnerability, as well as the labor, that bureaucracy generates for all involved means that in some instances it is sympathy rather than disdain that IFF staff display toward state officials. In addition to complicating the critique described in the previous chapter, attention to the variety of critical and sympathetic responses that emerge out of IFF staff's interactions with bureaucrats can help us understand what is at stake in IFF staff's efforts to enroll the participation of vulnerability-averse officials in research and development work.

Getting Screwed from Both Sides

Along with producing vulnerability, bureaucratic workarounds generate remainders in the form of further additional bureaucratic labor. This potential is particularly apparent in relation to the possibilities and the headaches created by the office's dual legal identity—its existence as both the country office of an international organization (IFF) and a center within a domestic public research institute (SAPS). This dual identity allows several advantages, including access to two pools of funding—one international and one domestic. Whereas most international research organizations cannot apply for domestic research grants—

including those from NSFC and MOST—IFF can do so through its persona as a SAPS center. The office's dual identity can also allow it to escape certain obligations to IFF's Africa-based international headquarters. For example, the China office is required to allocate 30 percent of any grant funds to IFF's overhead expenses. These overhead funds are intended to support indirect but crucial institutional expenses like offices, support staff, and laboratory facilities. Asking donors to pay overheads of this kind is a common practice, but 30 percent is unusually high, and budgeting such large overhead costs can lead to objections from donors who prefer a higher proportion of their funds to be spent on direct project expenses. In addition to the headache of negotiating high indirect costs with donors, some IFF staff members themselves object to this overhead policy. Even where grants are won specifically by IFF-China, the 30 percent of funds allocated to overheads rarely makes it to China and is instead swallowed up by IFF's global headquarters. This 30 percent allocation can, however, be avoided by applying for a grant as SAPS rather than as IFF. Where this is done, the grant becomes subject to the much more modest 5 percent overhead allocation required by SAPS.

The ability of the office to select which of its legal personas to adopt in any given situation gives it a degree of latitude in its administrative and bureaucratic dealings with both Chinese government offices and IFF's international bureaucracy. This leeway is in some respects similar to Gao Bingzhong's (2014) description of a religious committee in Hebei who overcame intractable government restrictions by establishing their new temple as at once a "temple" and a "museum." The Fan Village Dragon Tablet committee is devoted to the worship of a Dragon Tablet deity. Gao tells us that although there is increasing tolerance and even official promotion of folk culture, government policies continue to condemn deity worship as "feudal superstition" and to suppress its practice. As such, organizations like the Dragon Tablet committee are in constant fear of being shut down by the state. The committee has therefore been proactive in managing its image in the eyes of government bodies. Having reestablished Dragon Tablet worship in the 1990s, the Fan Village Dragon Tablet committee decided to construct a permanent temple structure. Building a temple, however, requires the permission of the Ministry of Religion: permission that, according to Gao, would have been impossible because Dragon Tablet worship is not an officially recognized religion. The committee's fix to this impossibility was to present its new structure to the authorities as a "cultural museum" rather than a "religious temple." This allowed the committee to avoid the Ministry of Religion's jurisdiction. At the same time, to raise funds for the building, the Dragon Tablet committee continued to tell the local community that it was building a temple. Thus, to the locals the building was presented as a temple, but to outsiders it was a museum. As with SAPS/IFF, the museum/temple dual identity generates a certain leeway in interactions with domestic (and, in the SAPS/IFF case, also international) bureaucracies, facilitating activities that would not otherwise have been possible.

Gao's ethnography of the museum/temple resonates with a broader anthropological interest in how communities sustain autonomy in the face of the totalizing claims of legal and bureaucratic institutions (Collmann 1988; Moore 1978; Pia 2016). Along with the latitude that bureaucratic tactics allow them, however, colleagues at IFF also highlight the bureaucratic work that such tactics can themselves generate. While seemingly a clever device to get the best of both worlds—the benefits of being an international organization as well as those of being a domestic public research institute—the office's dual identity also means, as Tao put it, "getting screwed from both sides." The cost of the dual identity is to become enmeshed in two cumbersome bureaucratic systems. In such a context, a workaround that allows colleagues to circumvent the constraints of one institution almost inevitably generates further bureaucratic headaches from the other. Running a grant through SAPS rather than IFF, for example, might save allocating 30 percent of grant funds as overheads, but it also subjects the office to SAPS's stringent anticorruption rules and procedures. These include especially cumbersome rules for international travel. Under these rules, a complete and detailed travel plan must be submitted at least one month prior to international travel—a particularly onerous requirement in the context of projects that routinely require plans to be made and changed at short notice. Conversely, if a grant is administered through IFF, there are fewer constraints on international travel, but there is greater scrutiny over whom the office enrolls as project partners. IFF procedures require all institutional project partners to submit themselves to an audit. This drawn-out procedure caused payments to two NGOs engaged in IFF's Agroforestry for Myanmar project to be delayed by nearly a year—a delay that almost led the NGOs to withdraw and the entire project to collapse. While creative bureaucratic tactics offer opportunities for mitigating or evading certain constraints or procedures, each tactic holds the potential to generate remainders not only in the form of vulnerability to future rebuke but also in the form of additional future administrative headaches and bureaucratic labor.

Anthropologists concerned with *guanxi* in China have highlighted the capacity for the exchange of gifts and favors to generate interpersonal obligations that extend into the future (Kipnis 1997; Yan 1996; Yang 1994). This capacity is apparent in the case of the Damei household survey described in the previous chapter. Having asked a favor on Kong Shifu's and my behalf from her classmate at the Damei Forestry Bureau, Professor Lin doubtless renewed a personal obligation to assist this classmate when he needs a favor of her. Similarly, on the completion of our survey in Meirong, Kong and I gave the village officials who had escorted us around the area a sum of "gas money"—what was in effect a cash gift politely disguised as reimbursement for transport costs. This gift was intended to reciprocate the assistance the official had given us, but this reciprocation was also done with the potential for future favors in mind. Though we had no plan to return ourselves, another group from IFF was due to visit the area quite soon, and we hoped that our generosity would predispose the village administration to our IFF colleagues.

Just as it is with gift exchange, the generation and control of effects that extend into the future is a crucial component of bureaucratic practice. Anticipating the immediate consequences and future remainders of a given bureaucratic transaction, as well as of the potential alternative transactions, is at the forefront of IFF staff's engagements with domestic and international bureaucracies. When IFF staff decide what procedural path to follow—whether to administer a grant through IFF headquarters or through SAPS, for example—their concern is with the future labor as well as the future vulnerability that alternative tactics might create. Whatever path they choose, IFF staff must frequently accept remainders and vulnerabilities that they would sooner evade. Nevertheless, navigating bureaucracy is, much like navigating informal social relationships, an art of anticipating and managing the potential future consequences one generates for oneself and for others.

Remainders as an Engine of Critique

IFF colleagues' frustrations with how bureaucracy generates vulnerability and unnecessary labor drive the critique of bureaucrats described in the previous chapter. The potential manageability of this proliferating labor and vulnerability can also precipitate a critical gaze toward their own colleagues. One of the most mundane examples of this is corporate services staff scrutiny of how closely researchers keep to relevant accounting requirements while on fieldwork or official travel. As described in the previous chapter, satisfying these requirements means collecting sufficient official receipts known as *fapiao* that are stamped by a business from the specific city or county one has visited. On returning to Songlin, research staff must give their receipts to colleagues in the corporate services team so that they can compile expense reports to SAPS. Accounting requirements often also entail a daily expenditure limit of around CN¥200 (US$30) per person plus transportation costs. Like the conference expenditure limits encountered by Tao's colleague, it is not always feasible to stay within these constraints. Indeed, overspending might be necessary because of high food and accommodation costs for IFF staff or because they must buy dinner and give gifts to local government officials helping them out. IFF corporate services staff are aware of these practicalities and are sympathetic to the difficulties that their colleagues face in the field. The corporate services team will, nevertheless, complain (though only sometimes openly) about research staff who they do not think have made adequate efforts to stick to the rules.

Over dinner one evening, Yanli, a member of IFF's corporate services team, shared her annoyance with a scientist who would always overspend and who did not appreciate the extra headaches this generated for her and her colleagues. This is not to say that the headaches this researcher creates are ever insurmountable. One possibility for administering excessive expenses would be to use the same name-borrowing device that Tao's friend used for her conference—to fabricate additional participants in the research trip in question. Another would be for IFF to cover excess travel expenses with discretionary funds that are not subject to such strict oversight. In Yanli's view, however, the challenges this colleague creates are in the first place entirely unnecessary. As such, her complaint focused the responsibility for resulting headaches not with the troublesome financial regulations or their authors but with the colleague who made no effort to work within the rules. Here, bureaucratic headaches bring into view not the irrational, arbitrary, or indifferent nature of the regulations involved (Gupta 2012; Herzfeld 1992) but the responsibility of colleagues to one another. In these instances, attention is brought to the externally imposed procedures that IFF must work within and around, but only to highlight the responsibility of colleagues to understand these procedures and to do what they can to minimize the headaches they foist on corporate services colleagues.

A more acute instance of corporate services staff's frustration with colleagues generating bureaucratic headaches emerged out of the failure of the office's communications officer, Wilhelm, to provide the appropriate documentation for a visa renewal. Wilhelm had moved to a new apartment several months prior to his visa expiring and had been told by corporate services staff that he must visit the local police station to register his new address. Registration with the police is a standard procedure required of all foreign visitors to China and something most international staff at IFF did themselves. Wilhelm refused to do this, telling a staff member named Yumei that she should simply use his old address on official forms or, failing that, go to the police station herself and register the new address on his behalf. This, unsurprisingly, angered Yumei and her colleagues to no end. This was, by all accounts, not the first time they had encountered entitled international staff asking them to deal with headaches that it was not their job to deal with. One thing that made this instance especially egregious, however, was that very few corporate services staff respected Wilhelm's work.

As one colleague explained to me, "When foreign scientists ask for assistance with things beyond our actual responsibilities, we often help them out. We know it is hard for foreigners to get by in China, and we know how important their expertise is to IFF. But I have no idea what Wilhelm does to make him so important to IFF." The motivation of IFF staff to deal with headaches correlates to the value (or lack thereof) of the work that overcoming that headache enables. With no one willing to do his bureaucratic work for him, Wilhelm decided to ignore the advice of his colleagues and not bother to register his new address with the authorities. Wilhelm subsequently gave the colleague responsible for his visa renewal documents showing his old address. Shortly after this, he departed China for an overseas holiday. While he was away, colleagues discovered that Wilhelm had knowingly given them documents showing an out-of-date address, but by the time they discovered this, they had already sent his visa application for preliminary processing with SAPS administrators.

The IFF colleagues involved did not share with me the details of exactly how SAPS administrators discovered and reacted to the deliberately incorrect address on Wilhelm's visa forms, but the response was enough for Wilhelm's employment to be terminated and for IFF's director to issue an angry warning to the office's foreign staff. At a weekly staff meeting shortly after Wilhelm's departure, Professor Yin demanded that foreign staff follow the corporate services team's instructions. "Do not challenge the system," he warned everyone, adding that IFF does not have a "magic wand" that would allow us to ignore the law.

Yin, moreover, emphasized the collective consequences of our individual actions: "If one staff member loses out on his visa, the whole office loses." Yin's fear here was that this one irregular visa application would bring the whole office under greater scrutiny and jeopardize all future visa applications. Indeed, Yumei blames Wilhelm for difficulties she subsequently encountered in procuring a visa for another colleague, Kathryn. Kathryn joined IFF shortly following Wilhelm's self-inflicted visa troubles. Kathryn had significant experience working with NGOs and international development organizations and was brought in to help IFF-China deliver on IFF's newly focused global mission to generate impactful research. Kathryn's expertise, however, fit less perfectly with SAPS—an organization dedicated to natural scientific research. This created a difficulty because it is through its legal persona as SAPS, rather than IFF, that the office can sponsor work permits. Justifying Kathryn as a SAPS employee was not in itself impossible. Indeed, Yumei had previously managed to secure a work visa for me—someone whose natural scientific credentials were no less lacking than Kathryn's. In the wake of the Wilhelm saga, however, Yumei told me that SAPS would be resistant to support IFF in anything but the most straightforward of visa applications. Yumei and her colleagues therefore concluded that it would be a waste of time to even attempt to get Kathryn a SAPS-sponsored visa. Instead, Yumei asked Kathryn to persuade her previous employer—an English language school—to continue to sponsor her visa.

This was a fix that did allow Kathryn to remain in China and to take up employment with IFF. It was, however, a fix that came at a cost and put Kathryn in a vulnerable position. Though gaming China's immigration system in this way is far from unusual, Kathryn was conscious that crackdowns on visa irregularities are not unheard of, and she bemoaned the fact that she did not enjoy the same legal status as her colleagues. Yumei and her colleagues blamed Wilhelm for Kathryn's situation. In their view, Wilhelm's actions reflected his lack of respect for the challenges that bureaucratic procedures present and by extension for the colleagues who would endure the labor and vulnerability that his careless actions might generate.

Self-Interested Bureaucrats

In addition to precipitating a critical gaze on colleagues, the impossibilities of Chinese bureaucracy can also generate sympathy for government officials. In one such case, Matt was having difficulty securing access to a field site in a national nature reserve for an international PhD student who wanted to collect data for the Asian Fungi project. This drawn-out process was somewhat frustrating, and at times infuriating, for Matt and his student. Discussing this case over lunch, Mubai pointed out that, as I had experienced in Damei, government officials can be sensitive to the presence of foreign researchers. Mubai suggested that regardless of the PhD student's intentions, if the official in question were to grant the student access to the field site, the official could find himself having to defend the decision to his superiors. The official might also fear, Mubai added, that responsibility for a foreign researcher in the reserve could harm his prospects for promotion. Expressing a degree of sympathy for this official, Mubai told me that if he were in the same position, he might take a similar course of action. Whereas Tao's complaint of SAPS administrators' refusing to trust entails a redistribution of vulnerability from bureaucrat to their client, in this occasion, Mubai recognized and sympathized with the vulnerability that a state official was stuck with. But if relationships with IFF threaten to exacerbate this vulnerability, how can collaboration ever happen? IFF colleagues offer two different answers to this question.

Mubai's sympathetic treatment of this government official is quite congruous with Yanli's previously described frustration with IFF colleagues. The obstructive nature reserve official and the IFF researcher who does not stick to the financial procedures both generate headaches for Mubai, Yanli, and their colleagues. But whereas the obligation to minimize headaches that Yanli imputes on IFF staff emerges out of their relationship to her as a colleague, the nature reserve official has no such relationship to IFF or its staff. Indeed, the obligation for an official to assist IFF is something that must be generated. In this regard, Mubai offered examples of how, in other instances, IFF staff had managed to cultivate the collaboration of interlocutors in government offices. He described how the willingness of Songlu Prefecture Forestry Bureau to collaborate with IFF in an earlier wild mushroom research project should be understood in the context of the funding that IFF brought to the project and IFF's agreement to make one of the bureau's staff a coauthor on project publications. By contrast to my experience in Damei, in Songlu the forestry bureau had an interest in placing itself in a relationship of responsibility to IFF and its work. Mubai further highlighted government policies that encourage collaboration between public research institutes and government bodies. By entering a collaboration, Mubai explained, both sides would be satisfying this policy. Mubai's point is simple: government officials collaborate with research institutions like IFF when it is in their self-interest to do so.

Mubai's analysis chimes with my own subsequent experience coordinating a collaborative experiment in matsutake mushroom management in Songlu. This experiment had emerged out of Frankie's ethnographic research with matsutake harvesters in a village named Boluo. She had observed some harvesters cover young mushrooms with piles of dead leaves or small sheets of plastic. Most harvesters reported that the coverings they use help mushrooms grow to their maximum potential, but they disagreed with each other about what kind of coverings work best. Frankie wanted to conduct a field experiment comparing the efficacy of various coverings. Seeing potential for this project to contribute to IFF-China's new outcomes-orientated focus, I accepted her invitation to collaborate, as did two junior mycologist colleagues. Boluo village would have been an obvious place to conduct the study, but Frankie had numerous reservations about running our experiment there, most especially her fear that Boluo's peculiar rotating forest tenure system would make it difficult to secure the continual access necessary for a multiyear study. As such, we instead took our experiment to nearby Songlu Prefecture. This is a place where we knew IFF had strong existing ties with the forestry bureau. Indeed, we hoped that we might piggyback on the same relationships that had facilitated the mushroom management experiment that Mubai referenced as an example of successful IFF–local government collaboration. That earlier project had been run by a senior IFF scientist, Professor Zheng, with whom I shared our proposal and asked for advice on working in Songlu. Zheng was generous enough to call ahead to introduce us to his contacts at Songlu Forestry Bureau, and he suggested to me that his previous collaborators there would be happy to help us out too.

When we arrived in Songlu, our first port of call was the forestry bureau, where we met Ms. Cui, an official who explained she had also worked on Zheng's earlier project. Cui invited our research team to sit down, and we walked her through our proposed experiment. Over the course of nearly an hour, Cui interrogated our plans and highlighted what she saw as the project's strengths and limitations, offering numerous suggestions for how we might improve the experimental design. Suddenly changing the subject in the middle of our conversation, Cui leaned in close and spoke in a lowered tone. The bureau's chief, she explained in a half whisper, was not at all happy with my colleagues and me. The bureau had been more than happy to collaborate with IFF on Zheng's earlier mushroom management research—a collaboration that involved her making weekly visits to the research site to monitor progress. But on that occasion, Cui continued, Zheng had funding to offer. On this occasion, by contrast, we had arrived expecting help for nothing. Having made it clear that we were pushing the limits of the bureau's hospitality and goodwill, Cui returned to a more hospitable tone and resumed discussion on the substance of our experimental design.

In this instance, Cui and her colleagues did not refuse to help, nor did they obstruct our work in the manner that Matt's PhD student had encountered. Cui accompanied us to our field site and introduced us to acquaintances in the local village who she thought might have an interest in our experiment. This was, however, our last interaction with Cui. She made it clear that we should not ask for assistance beyond that initial trip to Songlu. This encounter did not surprise Mubai when I related it to him on my return to Songlin. Indeed, my experience seemed to confirm his analysis that officials will only help IFF out when there is something in it for them.

For Mubai, none of this necessitates a moralizing critique. He is sympathetic to the reasons for officials' reluctance to cooperate with IFF. Mubai's sympathy for the government official who refuses to help IFF is perhaps an example of Mubai offering bureaucrats the kind of "ethical alibi" that Michael Herzfeld (1992) suggests bureaucrats often claim on their own behalf—the alibi that fault is with "the system," not the bureaucrat. Unlike Herzfeld's account of this ethical alibi, however, Mubai does not see it as entirely disingenuous. Mubai understands that officials are entangled in a system that is a source of vulnerability as well as of power. As such, Mubai offers an assessment of state officials that is somewhat more generous than Herzfeld's image of the dehumanizing bureaucrat (cf. Lea 2008; Pia 2017). Nevertheless, there remains a sense in which Herzfeld's and Mubai's accounts converge. By dichotomously opposing altruism to the self-interest and insisting it is the latter according to which officials act, Mubai's bureaucrat is in a certain respect just as indifferent as Herzfeld's. It is simply that in Mubai's account of IFF's collaboration with Songlu Forestry Bureau, we find bureaucratic self-interest happening to be productive of a public good.

Willfully Vulnerable Bureaucrats

The narrative of officials acting instrumentally and out of carefully calculated self-interest is by no means unusual in contemporary China (Li 2012; Zhang 2001; Zhang and McGhee 2017). It is, however, not an analysis shared by all of Mubai's colleagues at IFF—or at least not one that is applied by them all the time. Some colleagues offer explanations other than self-interest for why Chinese officials might collaborate with IFF.

During a two-day launch meeting that IFF hosted for the Lengshan Multi-Stakeholder Platform, attendees were led around numerous farms and plantations by Director Ma of the prefectural government's crop development office, followed closely at all times by a photographer from a local newspaper. During a debriefing the day after the meeting, Bob expressed his delight at Ma's enthusiasm, highlighting it as evidence that Ma could be relied on as a key partner for and champion of the project. Responding with a degree of cynicism, I challenged Bob's optimistic take. I highlighted Ma's eagerness for photo opportunities and the fact that Ma had not yet committed his office to any concrete activities. Echoing Mubai's self-interest-oriented analysis, I suggested that Ma's apparent enthusiasm might merely have reflected his desire to build his reputation via the newspaper publicity and his superficial association with an important-sounding international initiative.

Jiaolong, another member of the project team, challenged my assessment. She agreed that Ma had indeed used the launch and the invitation of journalists as an opportunity to gain face but argued that he was also motivated by a genuine desire to spur sustainable development in Lengshan. Jiaolong agreed that the journalist's presence had been organized to impress Ma's superiors. But rather than, as my naive comments had suggested, moralizing Ma's reputation building, Jiaolong recognized such work as a necessary aspect of any official's career. Like Mubai, Jiaolong thought it important to appreciate and sympathize with the precarious conditions that government officials must navigate. But for Jiaolong, there was more at stake for Ma than personal career advancement. Jiaolong explained to me that Ma had spent the past thirty years working to establish and build a crop development office for Lengshan. In fact, she told me, most prefectures do not have crop development offices, and it is only down to Ma's personal advocacy that Lengshan has one at all. Among the achievements Jiaolong attributed to this office was the establishment of an extensive experimental plantation where Ma and his team were exploring the potential of numerous new crop tree species for Lengshan. This vital resource is, according to Jiaolong, something that would not have existed without Ma's vision and determination to improve rural livelihoods in Lengshan.

For IFF, Jiaolong explained, the fruits of Ma's labors are crucial. A key goal of the philosophy informing the Lengshan Multi-Stakeholder Platform was to provide a space where new collaborative initiatives might emerge. At the same time, however, Jiaolong and her colleagues also hoped to use the platform as a vehicle for a parallel, preexisting IFF project: Eco-Friendly Rubber. This project would involve researching sustainable and environmentally friendly cultivation practices for rubber cultivation in Lengshan and neighboring Southeast Asia. A key pillar of this research would be in situ trials of various rubber agroforestry systems. Most rubber in Lengshan is cultivated as a monocrop or with a single intercrop like pineapple or coffee. The agroforestry systems that IFF scientists planned to trial in the Eco-Friendly Rubber project, by contrast, would include complex multicrop systems with as many as a dozen crop species cultivated within a single biologically diverse agroforestry system. The experimental plantation that Ma had established would be an invaluable resource for such trials. Through that plantation, Ma and his colleagues had provided not only preliminary research on an array of tree crops suitable to Lengshan's ecology and economy but also a place where IFF might procure saplings for its own experimental plots.

Unlike Mubai's analysis of Songlu forestry officials, the alignment of Ma's efforts and achievements with IFF's project goals was not simply the result of Ma pursuing a narrow self-interest. Ma's work to establish his office and its experimental plantation was driven, in Jialong's view, by a desire that they both shared: to develop diverse and sustainable cropping systems for Lengshan. Reproaching my inability to see beyond this bureaucrat's self-interest, Jiaolong highlighted how climbing the hazardous hierarchy of Chinese officialdom might not be Ma's only motivation. Jiaolong, we could say, imputed on Ma a Theory of Change that is, like hers, animated by Lengshan's social and ecological future.

The difference between Jiaolong's analysis of Ma and Mubai's of forestry bureau officials in Songlu can be understood not only in terms of the explanatory power of narrow self-interest but also in terms of vulnerability. Mubai and Jiaolong both recognize the vulnerability to which their bureaucratic interlocutors are subject. This recognition is perhaps in itself powerful. It allows Mubai to place officials in a more sympathetic, and more human, light. Even in Mubai's sympathetic understanding, however, the number one priority for most officials is to minimize the vulnerability they inevitably face. In Jiaolong's view, this was not the case with Ma. Indeed, Jiaolong's engagement with Ma can be described in contrast not only to Mubai's accounts of bureaucratic self-interest but also to Tao's experiences with SAPS administrators. Part of what frustrated Tao about SAPS officials' refusal to trust was that they do so regardless of the consequences for others. In effect, when I described Ma as being interested only in gaining face, I had imputed a similar reading on him: I had naively suggested that he was only interested in shoring up his own position as a powerful member of a bureaucratic elite. The reason Jiaolong castigated me was because she sees the potential for something beyond this. Without denying that Ma indulges the imperatives of survival and success within the bureaucratic system he inhabits, Jiaolong pointed to work that created at least as many hazards for him as it avoided. In pursuing the establishment and nurture of a government office dedicated to sustainable agricultural development, Ma had opened himself up to vulnerabilities that more self-serving officials might have evaded. His willingness to take on vulnerability for the sake of sustainable development in Lengshan gave Jiaolong hope that her collaboration with him might be worth pursuing. Ma seemed, in other words, a potentially trusting and trustworthy collaborator.

The point in this contrast between Ma and Tao's interlocutors is not between an official who is vulnerable and one who is not. The difference is rather in divergent orientations to the inevitability of vulnerability: an openness to it as against an effort to minimize it at all costs. As we saw in the previous chapter, bureaucrats' refusals to trust and their insistence on rigid bureaucratic proceduralism can be understood as an effort to deny the contingencies, hazards, and entanglements that characterize social relationships. I related this to anthropological analyses of the future entailments that people variously generate and evade through carefully choreographed forms of transaction. This is a perspective that emphasizes the inescapability of vulnerability to social life. Recent feminist scholarship shares this assumption of inescapability but complicates the unambiguously negative connotations we often associate with vulnerability. This scholarship suggests, moreover, that the embrace of vulnerability can be a vital political act (Butler 2015; Butler, Gambetti, and Sabsay 2016; Parreñas 2018). In this more ambivalent feminist rendering, vulnerability is "not just a condition that limits us but one that can enable us. . . . [V]ulnerability is a condition of openness, openness to being affected and affecting in turn" (Gilson 2011, 310). Invoking this ethics of vulnerability as a mode of resistance in the context of the Kurdish movement, Nükhet Sirman (2016, 191) describes instances of "vulnerability that the subject embraces willingly, ready for the destruction as well as the creation that it will bring." Here, to embrace vulnerability is to open oneself up to the unpredictability, interdependence, and creativity of human relationships. I am concerned with vulnerability in a very different context and as a quite different mode of action—I am not trying to characterize the actions of IFF staff or their interlocutors as resistance. Nevertheless, compared to the bureaucrats Tao admonished—ones who closed themselves off to entanglements with IFF at all costs—Ma's openness to the contingencies and hazards of pursuing sustainable rubber cultivation is suggestive of openness to vulnerability as an ethical and political act. Whereas this is an openness and a vulnerability that Jiaolong recognizes in the practice of a government interlocutor, the remainder of this book will turn to focus on forms of openness and interruption in which we might ground an ethic of vulnerability for scientific practice.

ACCELERATING, UPSCALING, DESKILLING

In preceding chapters, I have referred to Theory of Change as a bureaucratic practice—a nomenclature that imitates the way that IFF colleagues subsume Theory of Change–based planning and evaluation regimes into the same category as the challenges of navigating state bureaucracies. For proponents of Theory of Change, however, a more appropriate category might be "tool." During the two years I spent at IFF, I witnessed several visiting experts extol the virtues of one or another new tool that scientists might employ in their work. Theory of Change is in certain respects typical of contemporary research for development tools. A structured method for planning and evaluating impactful research, Theory of Change epitomizes a scalar logic wherein effective engagement with a small but well-targeted group of next-users aims to generate changes in the behavior of a much larger group of end-users. Theory of Change also reflects a similarly typical connection between this scalar logic and value for money.

In other respects, however, Theory of Change—or at least the version of it that Lesley introduced during the outcomes thinking workshop described in chapter 1—is somewhat less typical. At her workshop, Lesley expected IFF scientists to identify the key barriers to achieving IFF's vision for the region. This is a task that demands a knowledge of how next-user groups—government, media, businesses, and so on—operate across a significant part of Asia. In the face of objections that this was not a subject about which we had much understanding, Lesley encouraged workshop attendees not only to identify problems in next-user behavior but also to map the kinds of activities that we should pursue to induce changes in their practices. Lesley was asking soil biologists, anthropologists, ecologists, and mycologists, against their better judgment, to speculate about the behavior of people about whom they know very little. Keen to see the Theory of Change design process through to a provisional conclusion by the end of her two-day workshop, Lesley urged her IFF-China colleagues to suspend their reservations. Rather than "getting bogged down" with what we do not know, Lesley pushed us to complete our Theory of Change as best as we could. This rough-and-ready approach to planning is out of step with the kinds of methods—or "tools"—that other IFF-headquarters staff and international donors are urging IFF-China colleagues to adopt. Most especially, the impressionistic basis of Lesley's Theory of Change contrasts with the highly systematized quantitative methods that otherwise dominate international agricultural and environmental research organizations like IFF.

In this chapter, I outline some of the characteristics and tropes common to the research for development tools I encountered at IFF. Most especially, I highlight pervasive imperatives for quantifiable certainty, deskilling, speed, scale, and value for money. Comparing the logics of these tools with Lesley's outcomes thinking as well as with aspects of IFF's Qingshan Agroforestry project, I illuminate the hegemonic logics that define "impactful research." These comparisons provide, moreover, my starting point for an exploration of certain less-orthodox aspects of IFF colleagues' work.

Quantifiable Certainty

The divergence between Lesley's impressionistic Theory of Change and the systematicity of other research for development tools is especially apparent in contrast to the virtues that Brian (one of Lesley's colleagues at IFF headquarters) espouses of his "decision analysis" tool. On a visit to IFF's China office, Brian gave a talk introducing his latest work to develop and implement this tool. The first step of a decision analysis, Brian explained, is to identify a decision that policymakers are in the process of making. They might, for instance, be deliberating the question "What agroforestry system should we promote for rubber production in Lengshan Prefecture?" Comparing the various alternatives available to the policymakers in question, decision analysis provides a quantitative method for calculating which decision is most likely to have the best outcome. This method uses existing data along with Bayesian networks and calculations of probability to model alternative decisions and their likely impacts.

Central to this method are calculations of confidence in the model—the degree of certainty one has that the model is correctly predicting the impact of the alternative decisions under consideration. These calculations of confidence pinpoint the knowledge gaps that are responsible for any lack of certainty. Thus, a decision analysis not only produces policy advice but also identifies areas to target for further research. As Brian put it, "We can identify knowledge gaps that really limit decision-making and try to close them." In contrast to the impressionistic predictions of how activities might generate outcomes and impacts that characterized Lesley's workshop, Brian's decision analysis promises both quantifiable certainty in the best policies to pursue and knowledge of precisely what new knowledge would help us increase confidence in our analysis.

A significant input for a decision analysis model is discussions with relevant experts on various factors salient to the decisions under consideration. These experts might be asked, for example, the yield they would anticipate from a given crop or the market price that various crops might achieve over the coming years. Brian referred to such evaluations as "subjective beliefs." On first impression, such assessments do not seem so far from the off-the-cuff assessments that Lesley asked IFF staff to make as they put together the China office's Theory of Change. Using a formalized process of calibration and aggregation, however, decision analysis turns these subjective beliefs into quantified probabilities.[1] One component of this process is something called "calibration training"—a method Brian borrows from the business consultant Douglas Hubbard's (2014) *How to Measure Anything*. This method is based on the idea that most people are either consistently *over*confident or consistently *under*confident in the estimates and the predictions that they make. Calibration training teaches people whether they are overestimators or underestimators and trains them to become more adept at accurately assessing confidence in their knowledge. A well-calibrated expert will not only be able to tell you, "I think that the yield for crop X would be between A and B," they will also be able to tell you, "I am 90 percent confident that this estimate is correct." Such assessments, along with various other inputs, are fed into decision analysis's mathematical model and aggregated with a variety of data inputs to produce a probabilistic assessment of the future outcomes that a given decision would produce. Thus, while some of the inputs might initially appear just as impressionistic as the discussions that animated Lesley's Theory of Change workshop, the mathematically precise endpoint of decision analysis is a long way from the "working hypothesis" that we produced in Lesley's workshop. As Brian puts it, decision analysis allows scientists to persuade next-users "to make better quality decisions based on good science and available evidence, and to make decisions and preferences more transparent."

The faith that Brian displays in the capacity of decision analysis evinces a commonplace notion of technical decision-making: an ideal of policy as a value-neutral weighing-up of alternative options in terms of objective costs and benefits. As Arturo Escobar (1999, 385–86; also Ferguson 1994; Li 2007; Mitchell

2002) argues, this pretense to value neutrality and superiority fosters a view "of social life as a technical problem, as a matter of rational decision and management to be entrusted to that group of people—the development professions—whose specialized knowledge allegedly qualified them for the task." In decision analysis, this logic of certainty and value neutrality is closely tied to increasingly hegemonic conceptions of scientific rigor that privilege formal and quantitative methods (Adams 2016; Porter 1995). Quantitative outputs provide a basis, moreover, for calculating a financial value to research—a calculation that feeds into a wider trend for speaking of research and policy in business and economic terms. In this respect, Brian wants to assuage what he sees as scientists' misplaced reluctance to put a precise monetary value on their research. During his presentation at IFF-China, Brian assured his colleagues that decision analysis can scientifically calculate the economic value of a proposed piece of research. He went on to suggest that where a policy decision is between different land-use types or different cropping systems, decision analysis may be able to show that "there is a business case for agroforestry."

Here Brian echoes Alistair's ambition to draw on the monetizable results of M&E in an IFF business strategy.[2] One could also read decision analysis as the predictive counterpart to the retrospection of M&E and the ubiquitous randomized-control trial (RCT) (Adams 2016; Song 2020).[3] In rural development, RCTs are often used to measure the impact of a project on local incomes. In RCTs, baseline data is collected on local incomes prior to project initiation, and changes are then tracked as the project progresses. Mimicking a stereotypical laboratory experiment, RCTs require data collection both at project sites and at control sites that have been deliberately excluded from the development intervention under evaluation. Comparing changes in wealth at project sites with that at control sites, development practitioners can calculate the change in local household income attributable to the project. Further factoring in costs, one can calculate the project's "return on investment"—the monetary value of the project's impact minus its cost to implement. Decision analysis likewise entails the assessment of return on investment for a given development intervention, but it does so predictively. It could be used, for instance, to calculate an expected return on investment from promoting agroforestry systems in upland Myanmar or in Lengshan's rubber-dominated landscape.

Deskilling Social Science

A conviction in favor of quantitative methods is by no means unique to decision analysis. Another IFF headquarters–based scientist, Josiah, took his visit to IFF's China office as an opportunity to urge his colleagues to adopt a tool he and colleagues had developed and deployed in South Asia and East Africa. His "Toolkit for Agroecological Knowledge" (TAK), Josiah explained in a seminar for IFF-China colleagues, is a protocol for collecting and collating local knowledge of agricultural and forestry ecosystems. Central components of TAK include a questionnaire survey and a computer program that uses survey data to rank local trees according to various economic and ecological attributes. Data collected and analyzed with this tool would, for example, tell you which tree species provide the greatest benefits to soil quality and which ones produce the best livestock fodder. A key part of Josiah's case for why colleagues should adopt TAK is the superiority of his protocol over what he dismissed as "anecdotal methods" (cf. Adams 2013). To illustrate his point, Josiah offered an example of a project where a team of scientists and development professionals had been trying to promote the cultivation of a particular tree species. Local farmers were refusing to adopt the tree, but Josiah and his colleagues could not fathom why. Only after they implemented his tool did the reason for the farmers' refusals become apparent: "What TAK revealed," Josiah explained, "was that this tree causes significant soil erosion."

By "anecdotal" methods, Josiah referred to the forms of informal observation and unstructured conversation that often characterize agroforestry scientists' efforts to develop a picture of a project site. Such methods are familiar to the Qingshan Agroforestry project that I discuss later as well as, of course, to anthropological fieldwork. Initially, I was unable to comprehend why Josiah was so certain his tool was superior to an anecdotal approach. In the example Josiah gave of a tree causing soil erosion, an anthropologist is inclined to wonder whether what he dismissed as anecdotal methods could quite easily have reached the same finding as Josiah's tool. Surely, a simple conversation with the farmers about why they disliked the tree in question would have produced the same conclusions with half the effort? To some extent, Josiah's heightened confidence in his tool may simply reflect a belief—shared with Brian—that anything can and should be quantitatively measured. However, I later came to appreciate an important capacity of Josiah's formalism over what he called anecdotal methods: it anticipates unskilled workers.

Bob, a China-based IFF research fellow who had designed a research tool of his own, was quite explicit about the significance of this deskilling capacity. In a presentation that outlined his ongoing work with this tool, Bob echoed certain attributes of TAK. Bob described, for example, how a standardized protocol enables the tool's easy replication across multiple research sites. At lunch following the seminar, however, Bob emphasized a motivation for his tool that differed from Josiah's commitment to the inherent rigor of highly formal methods. Bob explained to me how he was originally inspired to develop his tool by the incompetence of an IFF social scientist who had implemented a household survey that was so poorly designed that the data it produced is almost entirely useless. This survey, Bob complained, was implemented by a colleague prior to his arrival at IFF, but he was now being asked to turn the swathes of data from this survey into high-quality publications. Bob lamented how difficult it was for him to generate meaningful findings from a survey that was designed without a clear set of research questions in mind. He suggested, moreover, that this case is indicative of a broader tendency in agricultural research for development. Without coherent, defined questions and research goals, many researchers just throw every question they can think of into the survey, expecting to somehow find coherence in the data after the fact. Bob attributed this failing to a lack of social science training. People are designing and implementing surveys despite lacking the expertise and the know-how necessary to do so. Bob's solution to this was simple: he developed a prepackaged social scientific protocol that requires very little skill to deploy; he developed a tool that "anyone" could use.

Though Bob's comments concerned relatively senior professionals, the idea of erasing skill from the research process is also vitally important for how IFF scientists engage low-paid survey enumerators. Much of IFF's social science research relies—like Bob's tool—on data from household survey instruments. To implement these surveys on large scales and in a timely manner, IFF ordinarily employs a team of survey enumerators to collect data. Drawn from a combination of junior members of IFF's corporate services team and the student bodies of local universities, enumerators are well educated but not necessarily trained in the social sciences. The minimal training they might need to implement the questionnaires is therefore provided by IFF, often in an intensive workshop that can last from a few hours to a couple of days.

One of my responsibilities as a research assistant at IFF was to design and implement a household survey on wild mushroom harvesting in the Greater Mekong region. To do this, I collaborated with Frankie—the same ecological anthropologist who Mubai had helped secure access to an ethnographic research site and with whom I subsequently worked on the matsutake management experiment. Designing the household survey, Frankie and I instinctively devised open-ended survey questions as starting points for conversations with mushroom harvesters. We soon discovered, however, that such an approach would be impossible if we were to employ the survey enumerators available to us. After running trials with the survey, we quickly realized that lacking skills in social scientific research, the enumerators we had recruited were inclined to record short answers and to ignore what we thought were significant aspects of what interviewees told them.

In one trial of the survey instrument, a harvester told me that he never managed to find enough wild mushrooms to be worth selling but that he would nevertheless go collecting "for fun." This instantly stood out to me as incongruous with rural development and natural resource management literature that focus on mushrooms as financial or subsistence resources but seldom as objects of leisure. Noting down what this respondent had described, I took his response as an entry point for a further discussion about the recreational forays into the forest he and his friends would take. Meanwhile, the enumerator whom we had hired to implement the same trial questionnaire, I discovered that evening, had received similar responses but had not recorded this data in his notes. He had seen no reason to consider this information interesting or to encourage his respondents to elaborate on what they had meant by harvesting for fun. Instead, our enumerator had simply recorded that such respondents seldom harvest mushrooms.

The obvious lesson—and the naive failing of Frankie's and my research design—was that the informal research methods we had envisaged require tacit knowledge: a knowledge that one takes for granted but that is not easily transmitted to others. Having read extensively on mushroom harvesting and, in Frankie's case, having carried out her own participant observation with commercial harvesters, we had developed a tacit understanding of what aspects of a conversation with a mushroom harvester are interesting and worth recording.

This is not to say that the skills of informal interviewing or of collecting data through open-ended conversations were beyond the abilities of the enumerators we had hired. Or that these enumerators were not talented people. Indeed, in ways that we did not anticipate, our enumerators brought skill sets and tacit knowledge on which the survey fundamentally depended. Their familiarity with the informal conventions of local government in rural China was, for example, invaluable to us as we navigated encounters with various government officers and community leaders. Nevertheless, developing the specific skill set for the open-ended interviews we had imagined would have required a lengthier learning process than we had time for. Having been puzzled by why Josiah would prefer a complex formal method to much simpler and ostensibly more straightforward informal or anecdotal methods, I came to appreciate that informal methods are in many respects the more cumbersome to implement. This is especially so if one is expected, as Frankie and I were, to complete research quickly and at scale. Instead of developing a highly skilled cohort of enumerators with a tacit knowledge of what is and is not an interesting response to a survey question, we formalized our survey, introducing more yes/no and multiple-choice questions. Here we employed a similar logic to Bob. We wanted to make the survey in such a way that eliminated the need for survey enumerators to acquire a novel skill set. We attempted to create an instrument that *anyone* could implement.

This deskilling of research is analogous to the process Tim Ingold (2000, 316) describes wherein "[t]echnology . . . appears to erase technique." Suggesting that this is a distinctively modern and capitalist process, Ingold describes technique as "tacit, subjective, context-dependent, practical 'knowledge how,' typically acquired through observation and imitation rather than formal verbal instruction." Here technique would equate to the kind of skills that we found our enumerators to be lacking. Ingold describes technique as an embedded form of action and contrasts this to "technology, which consists in a knowledge of objective principles of mechanical functioning, whose validity is completely independent both of the subjective identity of its human carriers and of the specific contexts of its application." Described in Ingold's terms, research tools and instruments dis-embed their users from research activities, doing away with tacit knowledge and externalizing the productive knowledge-making force of the tool from them.[4] Much like the demography surveys that Crystal Biruk (2018) describes, deskilled research resembles a Fordist assembly line.

Deskilling Soil Science

A variation of this deskilling phenomenon is evident in the case of decision analysis. During his visit to China, Brian encouraged IFF-China to find a young scholar who could be trained to use the tool. Brian advised that a background in economics might be useful. Implementing the tool, he explained, requires a user capable of manipulating quantitative data in a manner familiar to economic analysis. But training this one specialist technician would at the same time make it unnecessary for other scientists in the office to develop skills in decision analysis. One person would be tasked with implementing decision analysis for all projects run out of IFF's China office. Even this person, moreover, would not need the higher level of expertise that Brian and his collaborators had needed to design the tool. A broad range of skills remains necessary in the world of research for development, but tools are technologies for minimizing the instances and scope of skills (or, in Ingold's terms, technique) that it is necessary for individuals to acquire. This logic motivates the design of activities that can be replicated by a hastily trained cohort of low-skilled professionals.

Terry Rambo (2007) observes an analogous case of deskilling when he laments how rapid rural appraisal (RRA)—a widely deployed approach to rural research and development and predecessor to currently ubiquitous participatory rural appraisal (PRA)—has transformed in recent decades. Whereas today RRA and PRA tool kits are, much like Bob's tool, designed to be implemented by low-skilled practitioners with minimal experience of a specific locality, Rambo suggests that this was not originally the case (cf. Li 2007; Mosse 2005). According to Rambo (2007, 794–96), early practitioners of RRA "held advanced degrees . . . had long experience working in the northeast Thailand [where the RRAs he describes were being implemented] and already had a very good understanding of the local situation." These practitioners were, moreover, aware of the limits of their tools and would never imagine them "as a way of providing definitive answers to complex questions." By contrast, Rambo argues, the way the current generation of practitioners deploy RRA has created "the illusion that [RRA tools] offer a cheap and easy substitute for more intensive methods of data collection." Here Rambo observes not only a deskilling in rural research for development but also that this deskilling has been accompanied by a diminished awareness of the limits to the knowledge that these tools generate.

Survey tools and survey instruments that operate according to this deskilling logic are not limited to the social sciences. Early on in my time as a research assistant at IFF, I was asked to travel to a tropical area of southwest China to collect data as part of a trial for a new tool for assessing on-farm natural resource quality. This tool had been designed by a senior scientist and soil specialist at IFF headquarters named Carlos. Carlos had recruited staff from across several IFF offices in Asia, Africa, and Latin America to implement the same tool. There was no expectation that these enumerators would share Carlos's expertise in soil science. The tool was meant to allow anyone (including a novice such as myself) to conduct a natural resources assessment.[5] By implementing the tool in southwest China, I would provide data that would help Carlos test his tool's efficacy. His hope was that the tool would provide a quick and reliable measure of local natural resource degradation in a diversity of global agricultural and forestry settings.

The tool consists of a scorecard separated into six sections representing six key dimensions to what Carlos calls "natural resource integrity": landscape context; soil erosion; soil organic matter; soil nutrient availability; soil biological activity; and pests, weeds, and diseases. Each of these sections contains five measures to be scored on a scale of one to five. Many of the measures are scored based on conversations with local farmers. The enumerator is instructed, for example, to ask locals about the frequency of soil tillage and about the presence of pest problems. Other measures rely more on the assessment of the tool enumerator himself. Here he is instructed to assess the visual presence of soil erosion or the slope gradient of cultivated land. How I was meant to score these measures was, however, rarely self-evident, not least in relation to the measurement of "thickness of topsoil" and "soil aggregate stability."

The thickness of topsoil assessment presented several problems. One was that the tool's numerical scale merely asked me to rate thickness on a scale of good to bad (five being the best, one the worst). But how thick is good? And how thin is bad? For a soil scientist with many years of experience studying farms and plantations, grading topsoil from good to bad is perhaps intuitive, but it was not intuitive to an anthropologist. I raised this and several other questions with Matt, who was helping Carlos coordinate his study. Matt suggested an easy solution to my topsoil depth problem. He gave me a measuring stick to take with me to the field and gave me a series of discrete depths. According to this revised scoring system, any depth less than 10 millimeters scores one, and anything more than one 150 millimeters scores five. This fix mirrors Frankie's and my realization that we needed to give survey enumerators discrete yes/no answers to record.

In the case of soil aggregate stability, finding a fix would be less straightforward. Having taken an introductory soil science class at Cornell University in preparation for my field research, I already had an awareness of what soil aggregates are and why their stability might matter. Crudely speaking, an aggregate is a lump of soil that holds itself together. If you pick up a soil that contains no aggregates, it will run through your hands like sand. A soil with strong aggregates, by contrast, will break up into larger lumps, each of which will require a degree of force to disintegrate. Laboratory tests can quantifiably measure soil aggregate stability, but manually assessing aggregates in the field is something that soil scientists do regularly, and this is what Carlos wanted me to do for his natural resource assessment tool. This was something I had done in my soil science class and is in some respects quite straightforward. One digs up some soil and then breaks it up with one's hands to see how well the aggregates hold together. The problem, however, was that this is an extremely tactile skill and one that requires a memory of what aggregates of varying stabilities feel like when broken up.[6] I could run through the process of digging up the soil and breaking it up, but I did not possess a soil scientist's highly developed sense of what a relatively stable aggregate feels like. What Carlos was asking me to do was not in and of itself unrealistic. I have witnessed soil scientists—both during the class I took in the United States and at IFF—describing the aggregate stability of a soil they are encountering for the first time. Had a soil scientist been implementing Carlos's tool, she would have had no problem assessing whether a soil aggregate is stable or unstable. Carlos's error was to forget how much training, experience, and skill this kind of judgment requires. On this and other measures, he had failed to reduce natural resource assessment down to a simple transferable tool that could be used by an unskilled enumerator.

To be sure, this was not the only failing of Carlos's tool. But what is interesting about this failing is the way that it parallels the failure of Frankie's and my initial mushroom survey instrument and the way that it highlights the diversity of techniques and skills that research for development tools must erase to be effective. Just as Frankie and I had overlooked the tacit subject-matter knowledge and research skills that our questionnaire would require, Carlos's natural resource assessment tool underestimated the degree of skill and of well-honed feel necessary to assess soil aggregates. For his tool to work, Carlos might need to redesign his protocol so that—like Bob's tool and Frankie's and my revised survey instrument—it could function in the absence of skill.

Scale, Speed, and Value for Money

Tools like Josiah's and Bob's offer the possibility of replicating a standardized protocol on a large scale. Once developed, prefabricated tools can be quickly deployed after only a short formal training. The fact that minimal skill and tacit knowledge are necessary not only accelerates the training process but also expands the number of potential users. If anyone can employ these tools, then the potential number of users of these tools is increased and with it the potential scale at which they can be deployed. The impact of a tool designed by IFF is multiplied by its adoption and implementation by actors outside IFF. Here a tool might begin life as something deployed in a relatively narrow set of sites. In the case of Carlos's natural resource assessment tool, he had selected Lengshan and his other sites across Asia and Africa to correspond with the location of a series of multistakeholder platforms, all of which were sister projects to the Lengshan Multi-Stakeholder Platform. Carlos hoped that implementing the tool across these sites would inform the process of designing research and development interventions within each of these platforms. But Carlos also hoped that once its efficacy had been proven, he could subsequently "upscale" the tool to be employed across a much broader spectrum of projects and initiatives. In this way, easily used tools epitomize the scalar logic of impactful science.

This same scalar logic is embedded in Lesley's Theory of Change. When Lesley asked IFF colleagues to construct a set of outcomes that would help us bring our vision into being, she asked us to do so with the full geographic scope of IFF-China's office in mind. By this point, IFF-China had responsibility for projects not only across China but also parts of Central, South, and Southeast Asia. As we workshopped potential outcomes, Susanna objected that it could not make sense for her research on soil antibiotic residues in China to be working toward the same set of outcomes as another colleague's project on watershed conservation in Central Asia. Lesley insisted, however, that we must indeed have a single set of outcomes to which the project on Central Asian watersheds and the other on antibiotic pollution in Chinese soil are both oriented. As described in chapter 1, there is a significant temporal incongruity between IFF scientists' and Lesley's imaginations of research practice, but there is also an important spatial dimension to what Lesley was pushing. The idea of focusing—initially, at least—on a local or national level is perfectly justifiable within Lesley's Theory of Change framework. The key, however, is that local-level work should be motivated by a plan for how those activities will eventually contribute to changes that are significant at the level of IFF's vision *for the region*. If Susanna has no such plan in mind, then from the point of view of IFF's organizational Theory of Change, her research is poorly planned and poorly motivated. Once Lesley had dictated that IFF colleagues generate a Theory of Change at a regional level, that is the scale at which activities should ultimately anticipate impacts.

As well as maximizing the scale of impact, tools serve as an imperative to accelerate the pace of research for development by promising results within extremely short time frames. One example of this emphasis on speed is a tool developed by Dennis, a European researcher who visited China as part of the Lengshan Multi-Stakeholder Platform project. Dennis's tool brings together representatives of farming communities, research institutes, the private sector, and government in a two-day workshop. Participants collectively determine core barriers or problems in local forestry and agriculture and identify entry points for future activities. Dennis and Bob used this tool to run a workshop in Lengshan Prefecture. Through this workshop, Dennis and Bob could, in just forty-eight hours, identify water pollution as the most appropriate focus for IFF's future activities (though subsequent budget cuts meant nothing came of this plan). These rapid methods appeal to donor preferences for speed and cost minimization. As Terry Rambo (2007, 795) puts it, "Why fund an ethnologist to live in a village for a year when you can claim to understand everything by sponsoring a three-or-four-day rapid appraisal exercise?"[7]

Imperatives for scale and tempo are intricately tied to an imperative for research and development activities to provide value for money.[8] This is true in the sense that technologies like decision analysis or Bob's survey instrument can be used to generate monetized predictions or measurements of an intervention's impact and of its returns on donor investment. Relatedly and more importantly, the goals of speed and scale are themselves driven by their connection to value for money. This connection is central to the logics according to which Lesley's Theory of Change imagines next-users as the intermediaries through which IFF impacts more numerous end-users and according to which certain IFF scientists valorize speed and scalability as a core feature of tools. Value for money is a defining feature of contemporary agricultural research for development and of what it means for research to be impactful. What matters is not merely that one's research anticipates making a difference in the world but that it anticipates that this difference be of greater monetary value than the cost of the research itself.

Anxiety in Soil Training

Aspects of Theory of Change share logics of quantification, scale, and value for money with other development tools. At the same time, however, the version that Lesley introduced to IFF colleagues is out of sync with decision analysis's promises of quantifiable certainty. One of my initial reactions to Lesley's workshop was an aversion to its amateurism. It seemed somehow shocking that at a well-regarded research organization such as IFF the methods for knowing the social and political context in which we were working could be so crude. This sentiment was shared by some IFF colleagues. Participants in the workshop pushed back at Lesley's invitation to share our impressions of how next-users in the region operate, and Lesley had to persuade participants to persevere despite our collective ignorance and uncertainty. As I encountered the intense drive for professionalization and value for money at IFF, however, I came to appreciate Lesley's rough-and-ready approach to Theory of Change for the relief it provides from the excessively grand and precise pretenses of RCTs and decision analysis. The Qingshan Agroforestry project provides a more extensive, and perhaps more deliberate, instance of IFF colleagues pushing back against the logic of tools like decision analysis.

The Qingshan Agroforestry project's goal was to introduce ecologically complex, biologically diverse agroforestry to a region that has suffered significant deforestation and where monoculture cash crop plantations dominate the agricultural landscape. By the time of IFF's involvement in the project, the initiative had been going on for several years, and the landscape was already dotted with fruit and nut trees that were meant to serve as the foundation for the planned-for agroforestry systems. The project's sponsor was, however, unhappy with the progress. Metelli, a European cosmetics company, had been funding a compatriot NGO named Arboreus to promote agroforestry in Qingshan as part of its corporate social responsibility program. On a visit to Qingshan, staff from Metelli and Arboreus observed that many of the trees were in poor health and expressed concerns that the project was not yet on the path toward the ecologically complex and biologically diverse agroforestry landscapes that the project envisioned. Staff from Metelli and Arboreus believed that a major barrier to progress had been the local partner NGO's lack of agroforestry expertise. Though this NGO, Shehui, brought invaluable experience working with communities in the region, it had never run a project focused on agricultural or environmental sustainability. To plug this gap in expertise, IFF was asked to work alongside Shehui, bringing its agroforestry knowledge and experience to bear through, among other things, the delivery of training to farmers in Qingshan.

IFF's remit for the project was relatively informal. IFF had a lucrative contract with Metelli to facilitate laboratory research on local medicinal plants with the potential for use in cosmetics. Given the importance of this contract to IFF, when IFF offered its assistance to Metelli's Qingshan project, it asked only that Metelli cover basic travel expenses. In contrast to the Agroforestry for Myanmar project, the absence of a formal funding agreement meant that IFF was free of the usual predetermined deliverables and rigorous monitoring frameworks. An imperative for value for money was, moreover, largely absent. Indeed, the professional logic of value for money would have made the project an absurdity. The smallest training workshop involved three IFF scientists delivering training to a group of just five farmers plus Shehui's project manager and a representative from Arboreus. Even the largest workshop IFF ran in Qingshan had only twenty-five participants. In a gesture toward the hegemonic scalar logic, workshops were envisaged as "training of trainers" so that those in attendance were expected to share what they learned at the workshops with their immediate neighbors in Qingshan. Nevertheless, even if those at workshops disseminated the outputs of workshops as hoped, the participants represented only two villages and a few hundred households. This is nowhere near the kind of scales that Lesley imagines for her Theory of Change or that headquarters scientists imagine for their tools. At such a small scale, IFF colleagues could not justify their involvement in the project in terms of value for money, nor do they try.

With the project led by two foreign scientists—Matt and Susanna—I was drafted in to help facilitate workshops and trainings in Chinese. Though it falls outside the imperative for value for money, our initial impulse was somewhat conventional. Following the contemporary consensus in professional agricultural development, we decided to search for participatory tools that we could use in Qingshan. Our first port of call was a digital repository of training materials produced by IFF's international headquarters. Here we expected we might find a preexisting model that we could adapt to Qingshan. Despite the proliferation of research for development tools already discussed, we were unable to locate anything among IFF's multitudinous training materials that seemed appropriate to our ambitions for Qingshan. We therefore decided to design our own workshop formats from scratch. In doing this, we were driven by a loosely defined set of rationales. First, we decided that giving the community a say in the direction of the training workshops and the project more generally should be an intrinsic good. Second, we knew that the local ecological knowledge of

Qingshan farmers would be invaluable to understanding and enhancing soil management. Collaboration, in other words, would be necessary because we had little hope of planning effective agroforestry systems without the benefit of local farmers' expertise. Third, we anticipated that a participatory approach would enhance our chances of a positive response to the training. Such reasoning was by no means novel. These three rationales have a long history in rural development (Chambers 1983; Stirling 2005; cf. Cooke and Kothari 2001). But in committing to designing the format of our workshops from scratch, we nevertheless departed from certain of the logics of prefabricated tools. This departure includes approaching the project as a unique collection of activities with no anticipation of replication or upscaling. In contrast to the highly formalized protocol of a tool, moreover, we embraced a somewhat ad hoc approach—an approach that entailed a tentative faith that our skills and expertise would allow us to find a path as we went.

The format we arrived at for one of our first workshops began with us presenting a variety of soil management options, including mulching, intercropping, cover crops, and tillage techniques. Having described potential benefits and drawbacks to each strategy, we asked farmers to consider the potential of each option for their own farms, and I facilitated a dialogue between IFF's two soil scientists and the farmers. So, for example, Matt and Susanna had proposed that farmers should switch from their current practice of ploughing vertically up and down sloping farmland to ploughing horizontally across the slope. Farmers were initially unenthusiastic about this suggestion, responding that horizontal ploughing would increase soil erosion and that horizontal ploughing would also make it much harder for them to use plastic mulch. Susanna and Matt were unconvinced by farmers' claims that horizontal ploughing could increase soil erosion, arguing that experience elsewhere in the world shows that the opposite is true: horizontal ploughing decreases soil erosion compared to vertical ploughing. Following a lengthy exchange, Susanna and Matt managed to persuade the farmers attending the workshop of a compromise position wherein they might experiment with horizontal ploughing on a limited number of plots and wherein they begin those experiments with crops that would not require plastic mulch. Finally, at the end of the workshop, we split farmers into three groups and asked them to select the management options that they thought would have the most potential for their farms. Through this exercise, farmers selected three soil management options they would like to adopt. In response to these selections, Susanna proposed follow-up support that IFF could provide to help with the implementation, and we constructed the outlines of a plan for a further workshop and a soil management calendar.

Though it never occurred to us to put it in these terms nor to relate this experience back to Lesley's Theory of Change workshop, our off-the-cuff approach resonated with the mode of knowing and planning that Lesley had encouraged us to practice. The scope and scale of the Qingshan project are somewhat less grand than the regional Theory of Change that Lesley had asked us to develop, but it was no less a project that took IFF staff beyond the comfort of their core professional expertise and experience. As extensive as their knowledge of soils and agroforestry was, Matt and Susanna had never worked in this part of China before. This was, moreover, their first experience (of what they hoped would become many) running rural training and participatory development workshops. This novelty generated enormous uncertainty about how we should proceed. Echoing the guidance from Lesley's Theory of Change workshop, however, we did not let the limits of our knowledge and our uncertainty "bog us down." If we did not let uncertainty bog us down, however, it does not mean that uncertainty was erased. Nor does it mean that we brought this complexity under control by calculating and quantifying the uncertainties—the Qingshan Agroforestry project's approach was a long way from decision analysis. Rather, uncertainty remained a constant and incalculable presence.

This was a presence that precipitated anxiety: our uncertainty made us uneasy about the activities we were pursuing and the impact they might have. This anxiety extended not only to the methods through which we chose to engage Qingshan farmers—were we really generating a space for open dialogue or were farmers simply telling us what they thought we wanted to hear?—but also to the proposals we delivered. So, for example, one of the three novel soil management techniques that farmers expressed an interest in was ploughing fields directly after harvesting the previous crop (as opposed to directly before sowing the next crop). Susanna, who was behind this proposal, had described to farmers how this practice is employed in Europe to reduce winter soil moisture loss. Farmer enthusiasm for this strategy generated both delight and anxiety: delight that the workshop had provoked a positive response from the participants, but anxiety that the new strategy might not have the anticipated effect on soil moisture. Susanna was therefore at pains to stress to participants that while she expected to see similar effects to those observed in Europe, she could not be certain of this. Uncertainty in the effect that our intervention would produce precipitated an anxiety that did not bog down the project, but it did impose the incomplete and imperfect nature of our knowledge as an ever-present concern. Uncertain that the knowledge we offered would be of any use to Qingshan farmers and uncertain that we could generate a meaningful dialogue about this knowledge, our soil management training proceeded in the shadow of significant doubt.

In light of her uncertainty about the new practices she was proposing, Susanna suggested to farmers that they merely experiment with each of the novel strategies we introduced and observe whether the strategies had the desired effect. Beyond this, she also proposed returning to Qingshan to help farmers measure the comparative soil moisture of fields in which they had and had not implemented new soil management strategies. More broadly, we anticipated further workshops the following year during which Susanna and local farmers could review and discuss the successes and failures of newly adopted soil management strategies. The controlled comparison that Susanna suggested is in certain respects an ad hoc version of RCT. It sits, however, in stark contrast to Brian's decision analysis. If Susanna imagines a potential to retrospectively know the effects of her work, this does not erase the uncertainty and anxiety that capture the moment of this work being done, nor does it anticipate that data bringing closure in and of itself—whatever the data, there were further workshops and conversations to be held. This is a long way from decision analysis's promise to provide calculable certainty of the best and most impactful course of action right from the beginning. In contrast to research for development tools, the Qingshan project generated anxiety rather than quantifiable certainty. This is an anxiety, moreover, that created an urge to proliferate opportunities for local participants and local knowledge to interrupt and to shape the project's trajectory.

Scenes That Escape Scale, Speed, and Value for Money

Research for development tools constitute a diverse genre, but they often share several core interrelated aspirations: quantifiable certainty, the deskilling of data collection and project implementation, deployment at scale and speed, and value for money. These are characteristics that cut across divisions between the natural and the social sciences. Soil scientists and anthropologists possess very different sets of embodied skills and tacit knowledges—a sense for what is an interesting comment for a harvester to make about mushrooms is very different from the sense of what it feels like for a stable soil aggregate to break up in one's hands. Both, however, are skill sets that require training and experience to acquire and that tools seek to render unnecessary. A good research for development tool is one that can be used by an unskilled user. In making such unskilled use possible, the tool becomes available for rapid implementation and upscaling. Producing a tool is, in this respect, the epitome of value for money. The initial investment in its design and trial implementation promises a large return through its subsequent uptake by a multitude of next-users in myriad settings.

The Qingshan project sits in tension with these trends. This is not to say that the project was entirely out of place in rural development practice. In its embrace of a participatory philosophy, it was very much in line with a mainstream approach to rural development—an approach so widely adopted that critical scholars have described it as "tyrannical" (Cooke and Kothari 2001). Moreover, for all the rhetoric of hyperformalized research for development, I suspect that one could find an abundance of similarly off-script projects elsewhere in the world. Nevertheless, the Qingshan Agroforestry project draws into relief some of the peculiar logics that inform institutionalized demands for impactful research and provides an example of the desires many of us harbor for space to work outside those logics. Here, perhaps, IFF staff share with anthropologists an impulse "to create scenes that exceed or escape" the narrow imperatives that govern contemporary scientific and academic institutions (Tsing 2015, 285; also Grimshaw and Hart 1994). In the two chapters that follows, I consider some of the tacit knowledges and research skills that are suffocated by imperatives for value for money before turning to further consider Qingshan and other scenes where IFF staff seek to escape certain norms and demands of professional science. In doing so, I will explore the relationships that such scenes facilitate and foreclose.

A FEEL FOR THE ENVIRONMENT

My first experience of IFF field research was to join Matt and Kumar as they led four fresh-faced graduate students to visit a research site where one of the students would conduct fungal diversity surveys for the following three wet seasons. The site itself consisted of neatly marked grids that the designated student was to visit on a regular schedule to record the mushrooms that he found there. On the face of it, this setup typifies the hyperformal and quantitative emphasis that undergird the version of scientific rigor evoked by tools like TAK and decision analysis. The protocol would involve students returning to clearly marked grids on a strict schedule spread across three years. On each visit, they would carefully record the mushrooms they found inside the predetermined grids. Combining data from a series of similar plots across a range of representative landscapes across southwest China and mainland Southeast Asia, the project would produce a picture of the region's macrofungal diversity.

What excited Matt and Kumar as much as the formal protocol they would implement, however, was the hour or so of hiking through the forests that was necessary to get to the research site. This was a hike punctuated by constant interruptions from Matt and Kumar as they stopped us all to point out the wonders of the forest, from insect-devouring cordyceps mushrooms to fantastic beetles and unusual tree species. This fascination with forests and the enjoyment of walking them inspire many IFF scientists to pursue the work they do. As well as inspiring interest in forest ecosystems, this playful exploration is a productive, even essential, aspect of scientific training and knowledge production. This is certainly the case for Kumar, whose passion is macrofungal taxonomy. Kumar grew up in South Asia before moving to Thailand and then China to train and work as a mycologist. Known and admired by his colleagues for his encyclopedic knowledge of and fascination with mushrooms, Kumar had gone to great lengths to get where he was. As he explained to me on a flight home from one of numerous visits we shared to rural southwest China, Kumar had set his heart on studying with an internationally recognized mycologist named Boris at Chao Phraya University in Thailand. Despite Kumar's strong academic record, Boris initially rejected Kumar's application, telling him he had a policy of not working with South Asian students. Not willing to let Boris's racism get in the way of his ambitions, Kumar "pestered and pestered" him until he eventually relented. In the role he eventually took up at IFF, Kumar trained doctoral students in Matt's soil biology group to identify wild mushrooms and to collect data for the Asian Fungi project's survey of fungal diversity across the region. This survey, Kumar told me on the same flight home, is something he hopes one day to repeat in his own home country. "There are fantastic mushroom field guides for places like China and America," he told me, "but no one has ever made one for where I grew up. I want to make one."

For this taxonomic work, noticing unusual mushrooms in the forest is a vital skill, and as intuitive as it seemed to Kumar, this skill did not come easily to the anthropologist. For every dozen rare or undocumented species that he spotted on the numerous walks I made with him during my two years in southwest China, I would pick out one or two species so common that they had long ceased to interrupt Kumar's path through the forest. This was something for which Matt took great pleasure in teasing me, often asking me how I had managed to walk past the most fantastic plants and mushrooms without so much as noticing. There was no simple tool that would allow me to contribute to this ad hoc data collection. Just as I could not expect the enumerators for my household survey to rapidly acquire the intuition necessary to social science research, I could not mimic mushroom-observing skills that Kumar had taken years to hone. Without the time or the discipline to develop the skills that Kumar possessed, on walks in the forest, I contributed only callow questions.

Walking in forests is something that frequently animates weekends and time off just as much as journeys to and from field sites. The practice in attention to the environment that this play allows can feed directly into formal scientific research. When he later joined IFF as a postdoc, Kumar embarked on a project to cultivate several previously undomesticated mushroom species. This domestication process often begins with Kumar finding a wild specimen of the mushroom he wishes to cultivate. In the case of one valuable medicinal mushroom Kumar was searching for, Matt told Kumar that he thought he had seen this mushroom a year or so previously when he had been walking in a forest searching for potential rock-climbing crags. Some days later, Kumar and I followed Matt an hour or so into the forest to the location where he recalled seeing this mushroom. Sure enough, with some foraging, Matt found another specimen. In this instance, a leisurely walk in the forest produced an opportunity for Kumar's mushroom domestication research several years subsequently.

In addition to locating rare mushrooms, this kind of loosely structured curiosity can also be productive of novel research projects. On another occasion, I joined Matt and Susanna on a tour of Qingshan farms and agroforests. As Susanna dug up a clump of earth to get a feel for the kind of soils the farmers were working with, Matt spotted something surprising: a green root nodule on the surface of the soil. Ordinarily white or pink, root nodules contain bacteria that can fix atmospheric nitrogen, allowing the plant to which the nodule is attached to indirectly access gaseous nitrogen. This is part of a mutually symbiotic relationship in which bacteria provide nitrogen to the plant in exchange for sugars that the plant has produced through photosynthesis. Picking up the nodule, Matt explained to me that the green color indicated that the nodule was itself photosynthesizing. Remarking that there could be an interesting research project in this unusual finding, he speculated that this nodule might be fixing nitrogen for crops without extracting anything from the plant in return.

Coffee breaks at IFF are frequently animated by conversations about ideas for research projects like this that have emerged out of casual observations made during walks on farmlands and in forests as well as during conversations with local farmers and foresters. Imagined projects explored: the impact of livestock grazing on matsutake yields, the productivity of a rubber plantation where a forester had ditched pesticides in order to make his plantation "look like a rainforest," the ability of crickets to digest plastic, and the efficacy of various growth-promoting tactics employed by mushroom harvesters. The way that these project proposals emerged out of open-ended attention to the world reflects how environmental expertise is, as Sarah Vaughn (2017, 261) puts it, often born "out of the necessity of recognizing the power of the world to challenge and reshape expert paradigms." Some of these latent research proposals grew into fully fledged projects; others simply evaporated. Matt's photosynthesizing nodule research proposal was especially short-lived. Looking at the nodule Matt had found, Susanna pointed out that only a small fraction of root nodules would ever be close enough to the soil's surface for this phenomenon to be occurring on a significant scale. It was unlikely, therefore, that photosynthesizing root nodules could ever make a significant contribution to a crop's nutrient needs. The research Matt proposed, Susanna suggested, could not to be of much use to agricultural development. Matt responded with a lighthearted joke insinuating that she had bought too wholeheartedly into IFF headquarters' mandate for research to offer value for money through predefined, large-scale development impacts. Matt's joke nevertheless underlined how little hope his latent research proposal would have of garnering institutional support. Unlike a project to cultivate marketable mushrooms, there was no impact pathway Matt could draw from research on photosynthesizing root nodules to increases in the incomes of rural households.

Instances of unexpected learning and inspiration like stumbling on a green root nodule resonate with a common, perhaps cliché, vision of scientific creativity. To illustrate the significance of such creativity, Christopher Newfield (2003, 125–27) offers the archetypal example of how "messing around" throwing a plate in a Cornell cafeteria inspired physicist Richard Feynman. According to Feynman (quoted in Newfield 2003, 126), "As the plate went up in the air I saw it wobble, and I noticed the red medallion of Cornell on the plate going around. It was pretty obvious to me that the medallion went around faster than the wobbling." Feynman's subsequent investigations into these dynamics provided the foundation for research that won him a Nobel Prize.[1] As with Matt's stumbling on a mushroom or a photosynthesizing root nodule, a moment of messing around provided unexpected impetus for research. The potential productivity that inheres in such moments is rarely, and perhaps never, entirely apparent in the moment itself. Here then is a further reminder of the temporal incongruity at stake in audit frameworks such as MELA that impose predefined endpoints for scientific practice. As Newfield (2003) argues in the context of the twentieth-century university, new modes of managerialism undermine ostensibly ill-disciplined practices that are central to scientific creativity.

There is, however, an important point of difference between Feynman's messing around and the messing around that provides unexpected grist for research at IFF. When IFF scientists follow farmers across their land or take leisurely walks in forests, they are indulging interests that bear directly on their scientific work. In this respect, we might better describe what they are pursuing in these moments as a "playful mode of inquiry" (Dumit 2021, 102).[2] They are not, after all, merely messing with whatever inconsequential distraction they happen to have at hand. Matt, Kumar, and many of their colleagues' love of the outdoors and intrigue in the natural world inspires and informs passions for their research. When IFF staff seek opportunities in their work and in their spare time to explore the landscape, they are creating opportunities to be worked on by the beings and the environments that they study. The importance of this work recalls the advice of geneticist Barbara McClintock: "Over and over again, [McClintock] tells us one must have the time to look, the patience to 'hear what the material has to say to you,' the openness to 'let it come to you.' Above all, one must have 'a feeling for the organism'" (Keller 1983, 198; also see Myers and Dumit 2011; Puig de la

Bellacasa 2017; O'Reilly 2017; Yarrow 2019). When Kumar, Matt, and their colleagues wander forests and farms, they cultivate and exercise a feeling for the organisms and ecosystems that they study.

In addition to highlighting incongruity between bureaucratic evaluation regimes and scientific practice, moments of playful inquiry also remind us that the neatly marked grids to which I followed Kumar and Matt do not represent the only time and space that scientific practice occupies. If one observed Matt and Kumar only as they constrained their gaze to the macrofungi within the neatly prescribed grids to which we had walked that day in the forest or observed them only as they analyzed the spreadsheets of macrofungal data that this gazing produced, then one might easily reach the conclusion that science operates in "the a-temporal, rationalized and de-contextualized time of the clock" (Adam 1998, 388). Focus on walking in the forest or on time spent observing pet crickets, however, reminds us how environmental science is also punctuated by playful inquiry, by moments that both stand outside rationalized data collection and provide the impetus for this data collection.

As anthropologists, we make a virtue of the open-ended nature of anthropological fieldwork. At times, this is an open-endedness we juxtapose with the depoliticizing effects of rigid technical knowledge (Ferguson 1994; Li 2007; cf. Vaughn 2022; Yarrow and Venkatesan 2012). We should not, however, assume that the presence or absence of unstructured research differentiates anthropology from agricultural and environmental sciences. The difference lies instead in the sequence in which informal knowledge production relates to other aspects of the scientific enterprise. Matt noticed both the wild medicinal mushroom and the photosynthesizing root nodule in moments of playful inquiry that held the potential to give way to more structured data collection. In the latter instance, that potential quickly evaporated, but in the former, Matt's observations subsequently helped materialize a carefully controlled experiment in domesticating a wild mushroom. This structure stands in stark contrast to an anthropology in which the data collection is the most open-ended moment of our research—the moment when we are most open to the interruptions of the worlds that we study. Scientists at IFF are similarly reliant on the surprises that plants, fungi, creatures, people, and ecosystems can bring when they take time to be among them and to be interrupted by them. This playful openness is simply something that more often precedes and informs the moment of formally recorded data collection.

Of course, soil science is not merely ethnography with soil instead of people. Nevertheless, attention to the commonalities between anthropology and environmental sciences can help illuminate what is threatened by insistence that research provide value for money. This is not merely a case of marginalizing qualitative methods in favor of quantitative research (though this is surely happening too [Adams 2016; Porter 1995; Rottenburg et al. 2015]) but also a case of suffocating the unspoken informal and qualitative methods that have always undergirded the outwardly formal and quantitative sciences. In a certain respect, this informal work has always been erased. Scientific objectivity has long demanded that the scientist retrospectively erase himself in his research write-up, leaving only the mechanical functioning of a disembodied protocol (Shapin and Schaffer 1985). In a regime where value for money is a supreme virtue, however, these unspoken components of scientific practice are starved of institutional space. The same institutional logic that makes the rapid scalability of tools so appealing would also render the ill-disciplined and open-ended research I observed Matt and Kumar undertake a waste of time and money. Indeed, it is significant that ill-disciplined wandering in the forest is something Matt or Kumar must do either on their own time or in time seized on their way to their formal work. With the working day confined to research that can be tied to identifiable impacts and quantifiable return on investment, the vital labor of wandering forests is not merely invisible but also starved of time (Adams, Burke, and Whitmarsh 2014; Berg and Seeber 2016; Mountz et al. 2015; Stengers 2018). The challenge may, in this respect, be to find ways to protect and to cultivate practices that celebrate and care for—rather than obscure and suffocate—the array of intangible skills on which our knowledge production depends.

Devaluing "Nonscientific" Staff

Of course, the intangible skills and tacit knowledges on which the sciences depend do not belong only to scientists. In *Leviathan and the Air-Pump*, Steven Shapin and Simon Schaffer (1985) describe a series of air-pump experiments that went on to become a paradigmatic model for modern scientific practice. In these experiments, Robert Boyle choreographed his laboratory as a public arena in which gentlemen could witness the outcome of his air-pump experiments and could thus testify to the veracity of Boyle's claims. In his laboratory, however, it was only the gentlemen that Boyle's procedures constructed as witnesses to his experiments. When he wrote up his experimental findings, Boyle would include a list of those who had witnessed his experiment firsthand and who could thereby attest to the veracity of his findings. Even when women had been physically present in the laboratory, these lists would never include their names. As Donna Haraway puts it: "Women might watch a demonstration; they could not witness it" (1997, 31; citing Potter 2001). Likewise, the working-class men operating the bellows to Boyle's air-pump would watch the experiments to which their labor was indispensable, but they could not witness in the manner of Boyle and his gentlemen colleagues.

Along with facilitating the direct witnessing of his experiment in the first instance, Boyle adopted a "naked way of writing" through which gentleman colleagues might retrospectively witness experiments via the medium of a report. Through these reports, readers were invited to witness not the subjective performance of Boyle but the objective reality of the air-pump itself: "It is not I who say this; it is the machine," Boyle professed in his research reports (Shapin and Schaffer 1985, 77 quoting Boyle). The feat that Boyle achieved through these new scientific practices was to render himself "transparent, self-invisible," leaving only the objectivity of the experiment itself in view (Haraway 1997, 29). Though Boyle wrote himself out of his own experimental reports, the modest witness's "self-invisibility" is fabricated in a very different manner to the "simply invisible" bystander or bellows operator who can merely "gawk curiously" (Haraway 1997, 25, 29). The former makes himself invisible in the experimental scene so that he can constitute himself as the author of objective scientific discovery. The latter's simple invisibility, by contrast, renders her presence immaterial to the experiment and its findings.

Here the privileged authority of the gentleman witness is intimately tied to the systematic devaluation of class-based, racialized, and gendered others deemed unskilled and incapable of producing scientific knowledge (cf. Mavhunga 2018; Philip 2004; Raj 2007; White 2006): "Gentlemen's epistemological agency involved a special kind of transparency. Colored, sexed, and laboring persons still have to do a lot of work to become similarly transparent to count as objective, modest witnesses to the world rather than to their 'bias' or 'special interest.' To be the object of vision, rather than the 'modest,' self-invisible source of vision, is to be evacuated of agency" (Haraway 1997, 32). In this regard, in addition to considering how imperatives for speed, scale, and value for money threaten the intangible skills of scientists, we must also ask what tools and survey instruments make of the intangible skills and knowledge-making agency of the "anyone" who is called forth to implement them.

As described in the previous chapter, the "unskilled" work of implementing household survey instruments—a common component of tools—frequently falls to members of IFF's corporate services team. This was the case for the Changing Global Landscapes project (CGL). CGL's goal is to provide a common basis for tracking environmental change across the world's humid tropical ecosystems. In contrast to tools like TAK, the CGL instruments have a discrete and predefined set of research sites and are in this respect a slightly different genre to the tools described previously. Nevertheless, with research sites spread across three continents, the CGL survey instruments parallel a tool's highly formalized protocols as well as its pretense to rapid global application with minimal adaptation. Implementation of the China portion of the CGL's household survey was managed by Huizhen, an IFF administrative assistant. It is not an experience she looks back on fondly.

As Huizhen described it to me, a social scientist from IFF headquarters named John visited China to deliver a two-day session to train her and the enumerators on how to implement the instrument. Throughout this training, Huizhen and her colleagues raised numerous concerns that John repeatedly refused to engage with. There were, for example, several concerns around survey questions on land ownership. Huizhen explained to John that many of the categories of land tenure offered in his survey's multiple-choice questionnaire did not make sense for China. She told John that she suspected that for almost all survey respondents, enumerators would end up having to indicate "other" as the form of land tenure governing the respondent's agricultural land. She pointed out that this would obscure the variety of complex institutional arrangements under which households in rural China enjoy land tenure. Huizhen suggested a simple revision to the survey: "Why don't we add the names of common local tenure arrangements to the existing multiple-choice selections?" Huizhen also protested against asking respondents to describe the area of their farmland and the volumes of their harvests in metric units. Chinese farmers, as Huizhen explained to John, simply do not use the metric system. One by one, John cut down Huizhen's attempts to raise the glaring problems that she and the other trainees saw with the instrument. "We must stick to the existing protocol," John insisted over and over.

Another issue Huizhen and her fellow enumerators raised was John's insistence that they implement the survey using a tablet-based app. According to John's protocol, enumerators should read questions off the screen and then input answers directly into the tablet. Then at the end of the day, the data could be quickly uploaded to the cloud. Having practiced using the tablet during the training session, it was clear to all those with prior experience in the field that the software was too cumbersome. Not only was it time-consuming to scroll through the available options for multiple-choice questions, but the app also forced enumerators to ask questions in a predetermined order. As Huizhen explained, mechanically going through a series of questions one by one does not work: respondents quickly get frustrated with enumerators who approach a survey in this manner. Instead, Huizhen continued, one must follow the flow of the conversations that the interview questions provoke. If, for example, someone starts telling you about their children as an aside, a good enumerator does not simply ignore the aside and insist that respondent discuss whatever topic is next on the questionnaire. Instead, the enumerator skips forward to the demographic information section of the survey, records the information the respondent has already offered about their children, and asks further related questions—for example, about their children's education or about their grand-children. While this was common sense to Huizhen, she lamented that like so many other senior scientists, John had always relied on others to implement his household survey instruments. People like John, she suggested, have an "ideal-ized" conception of what survey interviews look like. Lacking Huizhen's ex-tensive experience and expertise, John apparently mistook the reporting of mechanical methods pioneered by Boyle for the reality of field research.

The upshot of John's notion of survey implementation as mechanical process was that enumerators like Huizhen must find their own creative ways to render an impossible protocol workable. This was something facilitated by the fact that John—like many in his position—would not accompany his enumerators to the field. To fix the inappropriateness of the metric system to local knowledge of a harvest, Huizhen asked the enumerators she supervised to simply record in lo-cal units and estimate metric equivalents later. So, for example, Huizhen and her colleagues determined a set metric weight equivalence for a "basket of mush-rooms." Despite John's insistence that they use his tablet software, moreover, Huizhen transposed the tablet survey into a Word document that she printed out so that her team could record responses on paper. Then, at the end of the day, the enumerators would copy the data from paper to the tablet. Another chal-lenge was with a focus group survey conducted alongside household surveys. This focus group entailed asking participants to draw a "resource map" of the local area. Huizhen quickly discovered, though apparently not much to her sur-prise, that her research subjects were reluctant to take pen to paper. Encounter-ing this problem in her first focus group, Huizhen simply improvised a new protocol: she would do the drawing but ask people to tell her what to draw. Whereas John's survey protocol was based on an idealized picture of what a conversation with a survey respondent would look like, Huizhen—much like anthropologists, mycologists, and soil scientists—relied on a well-honed feel for the world she researches.

Improvising solutions to units and multiple-choice options that make no sense in rural China, Huizhen performed a vital scientific labor: She rendered the Chinese data compatible with John's global data set. Creating such compa-rability is no small feat (Schinkel 2016; Stengers 2011). Indeed, generating data that can put disparate parts of the world on a single comparative plane is at the heart of the scientific value Bob and Josiah claim for their tools and of what IFF's senior scientists say CGL is meant to do. With the data Huizhen provided, John and his colleagues could compare the social dynamics of a rural community in southwest China with those of communities across the global humid tropics. They could do so, moreover, while acting as if the exact same questions had been asked of each of these communities and as if each community's response to a uniform set of questions created inherently comparable data points. Nevertheless, for Huizhen this process of improvisation was extremely frustrating, even infuriating. Advice and insight she had offered to help John adapt and revise his instrument for China was aggressively rebuffed. Huizhen knew, even if John refused to accept it, that they might have collected far more meaningful data if he had engaged his enumerators in the process of designing the survey instrument.

Huizhen's improvised approach to achieving compatibility across data sets allows John to exploit her tacit knowledge, skill, and expertise while simultaneously evacuating her agency as a knowledge producer. Demanding that Huizhen do the impossible and apply the instrument unmodified, Huizhen was forced to both modify the protocol and pretend that she had not done so. Through this procedure, John sustained the illusion of the mechanically uniform survey implementation on which his gentlemanly self-invisibility relies while simultaneously rendering Huizhen simply invisible.

Caring for Skill, Caring for Colleagues

As Shapin and Schaffer (1985, 281) describe, Boyle's paradigmatic air-pump experiments depended fundamentally on "contingent acts of judgement . . . [on] the transmission of pump-building and pump-operating skills." The production of Boyle as a modest witness, however, depended equally fundamentally on the masking of that contingent labor. This masking serves a fiction that continues to undergird claims to objectivity at IFF and elsewhere: a fiction that scientific data is the product of mechanical labor. Of course, most agricultural and environmental scientists know this to be fiction. They know that the successful design and implementation of experiments and field studies are contingent on tacit knowledge, skill, play, and a feel for their subject matter that are not described or even mentioned in research papers. Driven to satisfy demands for speed, scale, and value for money, however, the designers of research tools and large-scale survey instruments often pretend to turn that fiction into reality by providing research methods that truly require no skill—protocols that can be implemented by anyone. In a certain respect, this seems a democratizing move. These tools are designed so that anyone and everyone could use them. In practice, however, tools and survey instruments often remake familiar hierarchies. In the case of CGL, a well-paid and predominantly White male cohort of scientists enjoyed the authority of self-invisibility, while a predominantly Asian and female cohort of enumerators were made simply invisible and rendered immaterial to the scientific knowledge that they helped produce.

Of course, demands for value for money through rapid, scalable research can undermine and erase the knowledge-making skills of scientists and faculty just as much as those of a tool's presumptive technicians. The virtues of speed, scale, and value for money cramp space for Kumar's mushroom-hunting skills just as they do for Huizhen's feel for a household survey interview. In this respect, deskilling at IFF illuminates a need for conscious commitment to something that has long existed across the sciences and humanities but that seems increasingly precarious: the ability and time to care for the skills on which our research depends. Considering the enduring importance of care for research practices alongside the division of labor at IFF and in the academy more broadly can help us to recognize that what is at stake is not merely the epistemic labor of scientists but also that of the staff with whom we share institutional homes. The playful modes of inquiry through which Matt and Kumar subject themselves to the interruptions of the worlds they study are a vital component of scientific practice, but they are not the sole province of scientists. One of this book's arguments is that the future of sciences and of universities should be built on the proliferation of opportunities for us to be interrupted by the environments and the publics we pretend to understand and transform. This academic "us" must be one that includes and values the Huizhens just as much as it does the Matts and Kumars.

GENERATING AND EVADING VULNERABILITY

In chapter 5, I described the Qingshan Agroforestry project diverging from the dominant logics of impactful research. In this chapter, I explore a further sensibility that sits outside IFF's professional norms: desires to bring personal values and narratives into scientific practice and development activities. What this turn to the personal shares with the productive anxiety of the Qingshan project is the potential to make science vulnerable to the knowledges, values, and interruptions of nonscientists. Both entail forms of openness that disrupt the hubris and certainty of strategies like brand building or of tools such as decision analysis. In this chapter, I also further consider the idea that Qingshan Agroforestry provides an escape from the institutional demands to which IFF colleagues are ordinarily subject. As tempting as it is to embrace the romance of such an escape, I explore some implications and repercussions of escaping vulnerability to professional communities and to those who fund scientific work. Drawing out the tension between IFF colleagues' impulses to generate vulnerability in certain spaces and their desires to evade it elsewhere, I highlight challenges that span the professional worlds of anthropology and agroforestry.

Making Science Personal

At an internal meeting for the Eco-Friendly Rubber project, Chad, Jiaolong, and I discussed how Jiaolong would present plans for an in-situ rubber agroforestry experiment to a potential new NGO partner. Ultimately Chad intended to develop a working paper setting out the scientific and development case for their experiment and the project as a whole. But for now, a key part of this pitch was a PowerPoint slide depicting a spectrum of agroforestry systems from least to most complex. At one end was a rubber plantation with a pineapple inter-crop. This common cropping system is, in Chad's view, only a mild improve-ment on the bare earth that intersperses monocrop plantations. This simple agroforestry system still depends on chemical inputs and still provides very little habitat for flora and fauna. At the other end of Chad's spectrum is a mul-tispecies "analog forest"—an agroforestry system that mimics the structure and functions of a natural forest, incorporating a rich undergrowth and multi-ple species in the tree canopy. Chad and his colleagues argue that such systems offer numerous advantages for soil health, nutrient cycling, and pest control. Explicitly contrasted against the universalizable, industrially disciplined, and chemical-intensive designs that have characterized so many modernist agricul-ture and forestry initiatives, Eco-Friendly Rubber would harness the efficacy of forms found in existing ecologies (Kohn 2013; Vaughn 2017).[1]

Toward the end of this planning meeting, Chad raised an example of a rubber plantation he had recently visited in Thailand. Here, the farmer had attempted to recreate natural jungle vegetation beneath the canopy of his rubber trees. This was a model very similar to the biologically diverse agroforestry systems Chad was proposing for Lengshan. Reflecting on what Chad had observed, the three of us discussed the potential ecological, as well as recreational and aesthetic, value of agroforestry systems that mimic the diversity of natural systems. During the ensuing conversation, Jiaolong proposed making a video of this Thai rubber plantation to show Lengshan farmers. Chad was enthusiastic about this idea, imagining that local farmers might be impressed by an informational film. As the conversation meandered, Jiaolong also took the opportunity to raise a further proposal. She suggested that the Eco-Friendly Rubber project should "focus on more than just the economic" and proposed that we promote an idea of a "beau-tiful Lengshan"—an idea that the industrialized landscape of monoculture rubber plantations could once again be made beautiful. As part of this, Jiaolong, who is herself an amateur artist, suggested that we should recruit local artists to promote the theme through their artwork. Chad was quick to dismiss this idea as "a little tangential." In contrast to working papers, PowerPoint slides, and in-formational videos, Chad saw artwork as somewhat out of place in a research for development project, and he rejected the idea.

At a later meeting for the project, Jiaolong raised a similarly out-of-place idea. This meeting was conducted with project partners from local government and research organizations. We began with Jiaolong presenting plans for the in-tercropping trial on behalf of Chad—who was not present. Following the presentation, a scientist from a local research institute suggested that each participant share his or her own vision for ecologically sustainable rubber in Lengshan. During the discussion that followed, Jiaolong proposed coverage on local television as a medium for promoting interest in Eco-Friendly Rubber. Responding to this, Mr. Wu, a local government officer, suggested that he could provide some of the reports his office had prepared on rubber plantations as content for the proposed television coverage. Jiaolong corrected him. She had not meant, she explained, "something so cold" as government reports. Rather, she wanted people like those assembled at the meeting telling viewers about their work, what it means to them personally, and why they do it. Jiaolong's idea was neither actively endorsed nor dismissed, but evidently uncomfortable with Jiaolong's idea, the other participants in the meeting changed the topic and moved the conversation quickly on. Just as her proposal for artwork to promote a beautiful Lengshan had jarred with Chad's expectations, there was an obvious (and, from Jialong's point of view, quite deliberate) incongruity between this proposal for personal narratives and the conventional expectations of project partners like Wu.

Though Jiaolong's efforts to bring art into the Eco-Friendly Rubber project were frustrated, there are other spaces in which art does enter IFF's work. In parallel with a conference that IFF hosted in Songlin, staff co-organized a film festival. The first film shown, *Yak Dung* (Lanzhe 2010), was made by a Tibetan elementary school teacher with the support of a local NGO. The film depicted the myriad things that members of a small Tibetan community construct with yak dung—from courtyard walls and ovens to dog kennels and children's sleds. Another of the films (McLeod 2013) likewise focused on indigenous ecological knowledge. Describing the spiritual nature of the Altai Mountains of Central Asia, this documentary lamented a planned gas pipeline through the mountains, claiming that in failing to respect the spirituality of Altai, this pipeline would precipitate earthquakes and other disasters.

To some in the audience, these films had very little value. In the question-and-answer session, one conference participant objected that we should not let indigenous knowledge "get in the way of progress." He argued that, unless we could use science to support the Altai claim that a gas pipe would cause earthquakes, then we should not stop the gas pipe running through the Altai Mountains. The somewhat diplomatic response of Wyatt, an IFF scientist who worked in Altai and who had introduced the film, was that although science has brought us a lot of progress and "liberated us from superstition," there remains "something missing." He explained that we were hurtling toward destruction and that people like those in Altai gave us some alternatives. Rejecting voices that would say we can sacrifice things like the Altai Mountains, Wyatt shared his vision for a future that values sacredness, that creates space for what he says is missing from conventional approaches to development and sustainability. As Wyatt saw it, "Concepts like biodiversity have a utility, but there are things at stake here that cannot be scientifically measured." Altai's spiritualism, in this respect, provides a supplement to the cold utilitarianism of research for development.

Julie, another conference attendee, raised her hand to respond to Wyatt's comment. Julie suggested that "we"—by which she meant the multinational and multidisciplinary cohort of scientists and development professionals attending the conference—already recognize the sacred by being in awe of the environments we work in.[2] In apparent agreement with Wyatt's sentiments on the limits of scientific concepts like biodiversity, Julie argued that we do not have to quantify everything to appreciate it. She agreed with Wyatt, moreover, that we can learn from the ways that indigenous people know and revere nature. She also insisted, however, that we already have our own capacity to revere nature. She argued that it is because we recognize the beauty and wonder of the natural world that we work in environmental research in the first place. Rather than locate Wyatt's "something missing" in an unfamiliar realm, Julie suggested that this something is already a deeply familiar part of scientists' personal experiences of the ecosystems that we live and work in.

In his ethnography of an animal rights charity, Adam Reed (2017) describes how novel forms of managerialism have precipitated an ethic of professionalism akin to classical models of bureaucratic impersonality. Whereas personal commitment to the cause of animal rights was once a vital moral marker for charity workers, professionalism has now become synonymous with an "organizational ethics centered on the separation of competency or task from individual moral conscience" (Reed 2017, 169). If this is a model shared by classical forms of bureaucracy and contemporary managerialism, it also has an affinity to the impersonality of scientific knowledge as characterized by the notion of science as a view from nowhere (Weber 1946; cf. Haraway 1988). According to this understanding of science, a scientific standpoint sits in stark contradistinction to a personal one. It is in the context of such a taken-for-granted distinction that Julie must remind her colleagues that they already know how to be in awe of nature. It is, similarly, in the context of this distinction that Jiaolong's call for personal stories from scientists or government officials about their motivations for transforming Lengshan's ecological landscape were so out of place. Echoing feminist STS (Keller 1985; Haraway 1991; Hubbard 1988; Roy 2018; Subramaniam and Willey 2017a), Julie and Jiaolong tried to disrupt expectations that scientists speak in an exclusively impersonal voice.

The incongruity of Jiaolong's proposals with the professional expectations of Chad and Wu stems also from her departure from a narrow conception of evidence-based agro-environmental policy and advocacy. This incongruity is especially evident when compared with the marketing approach taken by Alistair.[3] Outlining his strategy for marketing and branding agroforestry, Alistair had explained that IFF would need to target each of its potential customers with focused evidence on how agroforestry could improve something about which that specific customer was concerned. As an example, he suggested that when promoting participatory approaches to government stakeholders, IFF should collect data to show that farmer participation in project design leads to improved uptake from farmers. Following the presentation that Alistair gave to his colleagues at IFF-China, I suggested—in a manner similar to Jiaolong's contributions to Eco-Friendly Rubber project meetings—that in addition to dealing in evidence, agroforestry scientists are driven by ethics and values, and that this might add a dimension to his agroforestry marketing strategy. Alistair refuted this suggestion, insisting that "science deals in evidence" and that our strategy must remain a matter of "evidence-based lobbying." Like the conference participant who only sees value in the Altai people's knowledge if it can be substantiated by science, Alistair stands behind the unique primacy of scientific evidence.

Alistair's categorical rejection of anything but evidence-based argument is a classic example of "boundary work" (Gieryn 1983; Jasanoff 1987; Lei 1999; Zhan 2001). The authority of IFF's expertise relies on demarcating evidence-based scientific knowledge from things such as opinion, ethics, and emotion that Alistair simultaneously constructed as less valid. As sociological studies of expertise and professionalization have shown, moreover, the exclusive authority professionals demand within their field is tied to an awareness of their profession's limited competence and to an understanding that a good professional does not transgress those limits (Neal and Morgan 2000; Reed 1996; Wilensky 1964; cf. Hall and Sanders 2015). In this respect, one could understand Alistair's impulse to maintain the appropriate boundary of professional work—evidence-based argument—as a strategy for sustaining IFF as one of those institutions entrusted to do the technical work of development (Escobar 1995; Ferguson 1994; Li 2007). By delimiting the sphere in which he permits scientists to speak, Alistair reinforces a claim to science's privileged authority within that sphere.

Jiaolong's proposals for art and for personal stories in research for development stand in contrast to the notion of professionalism as a limited but authoritative sphere of expertise. Rather than claim and then bound herself to a distinct realm of professional expertise, Jiaolong plays with modes of communication beyond our conventional imagination of the scientific sphere—beyond the sphere in which she might claim privileged authority. Julie's reminder that scientists know how to be in awe of nature is similarly discordant with the intense quantification imperative evinced by tools such as decision analysis. Jiaolong and

Julie both refuse a narrow boundary for evidence-based lobbying and the hubris that this boundary might support. Generating humility in the sciences will, as many have highlighted, involve scientists recognizing the limits of their knowledge (Farmer 2016; Jasanoff 2003; Taylor 2003; cf. Cho and Kim 2022). Jiaolong and Julie show us how refusing the limits to professional domains defined by their impersonality might serve as an equally important foundation for humility. Embracing the personal may, in this respect, be one way to make willfully vulnerable the certainty and exclusive authority that colleagues like Brian and Alistair too often claim for the sciences.

Oppressive Bureaucracies and Communities of Critics

There is a loose parallel between Jiaolong's frustrated desire to bring artwork and personal values into IFF's work and the off-the-cuff approach of the Qingshan Agroforestry project. In differing ways, they each disrupt predominant logics of research for development. Tools like decision analysis promise certainty—a technical answer to the path one should follow—and evince hubristic attachments to value-neutral knowledge and privileged expert authority. The approaches that the Qingshan team, Jiaolong, and Julie pursued, by contrast, point to potentials for scientific practices that are distinctly uncertain, that are value-laden, and that consequently temper the authority of scientists' expertise. Compared to the hyper-rational imaginations that underlie decision analysis, brand building, and many of the professionalizing demands to which IFF staff are increasingly subjected, it is easy to appreciate the virtues of the vulnerability that turning to the personal or going off the cuff might generate for scientific knowledge. There is also a sense, however, in which the Qingshan project team transgressing the usual boundaries of their professional expertise entails the project team evading forms of vulnerability to which their work is otherwise subject.

The substance of the Qingshan Agroforestry project's activities required IFF's project team to work beyond the realms of their professional expertise. Indeed, though an initial farmer workshop was supported by Professor Zheng, an IFF social scientist and veteran of numerous participatory projects, subsequent training workshops were run by a much less experienced team. Two members of this team—Susanna and Matt—were soil scientists; I was the third. As expert as Susanna and Matt were in the ecology of the kind of agroforestry systems the project promoted in Qingshan, our roles at IFF had not previously extended to running workshops for farmers. Of course, lack of expertise is, from a certain point of view, no hindrance to rural development work. Development tools are after all designed with inexpert users in mind. In this project, however, we had declined to implement a prefabricated tool and had instead taken an ad hoc approach. As described in chapter 5, our stepping outside the well-worn path of prepackaged research for development tools generated a form of anxiety and uncertainty that was productive of a persistent desire to invite local farmers to interrupt our plans for the project. Another implication of our rejection of tools and their logics, however, was a refusal to participate in the community of development experts whose logics we rejected.

In many instances IFF colleagues welcome the interruption of peer review for its potential to positively transform the future direction of a research project or paper. They value it, moreover, for its role in upholding the norms and standards of the disciplinary communities to which they belong. In Qingshan Agroforestry, by contrast, IFF colleagues stepped outside the bounds of the disciplinary communities to which they are ordinarily accountable. In this project, there was no obvious peer community to whose standards we might be held. This situation was perhaps analogous to what Elizabeth Hall and Todd Sanders describe of human dimensions to climate change research. Hall and Sanders (2015, 448; also Strathern 2006) argue that institutional encouragement to transgress "professional, epistemic, and methodological boundaries . . . [has] reworked and even dismantled key accountability relations and mechanisms." They suggest that emerging models of transdisciplinary knowledge undermine the traditional mechanisms of peer accountability that operate within scientific disciplines.[4] Our plans for Qingshan likewise anticipated no peer community who might evaluate and interrupt our work. This rejection of research for development professionals as our "community of critics" (Strathern 2006) was intertwined with the partial escape from the dominant logics of scale, speed, and value for money that this project provided. These are after all logics pushed not only by donor institutions but also by IFF colleagues like Brian and Alistair.

Anticipating Anna Tsing's (2015, 285) sentiment that we must "create scenes that exceed or escape 'professionalization,'" Anna Grimshaw and Keith Hart (1994, 254) earlier described "the oppressive bureaucracy of our jobs" brought about by anthropology's professionalization. This is a professionalization for which they suggest we should blame Bronislaw Malinowski. They ponder, moreover, the alternative paths anthropology might take if we were to reinvigorate the amateur ethos of the "free-thinking intellectual" that they suggest died with Malinowski's predecessor W. H. R. Rivers (Grimshaw and Hart 1994, 237). The notion of the free-thinking amateur certainly has a heightened appeal in the context of the oppressive bureaucratization of the academy that Grimshaw and

Hart describe. The appeal is perhaps stronger still if we follow Edward Said (1994, 82) in defining amateurism as "an activity that is fueled by care and affection rather than by profit and selfish, narrow specialization."

There is something of an amateur spirit in the Qingshan Agroforestry project: both in its escape from oppressive, bureaucratically policed mandates and in the care and affection for Qingshan's people and landscape that motivated the project team. As I described in chapter 5, Susanna and the Qingshan team used their freedom from bureaucracy not to operate autonomously but to proliferate opportunities for Qingshan farmers to interrupt their work. In this respect, escape from bureaucracy facilitated not so much free-thinking as an opportunity for vulnerable thinking. As well as further highlighting the significance of this vulnerability and the kinds of escape on which it relied, in the remains of this chapter, I want to follow through on Erinn Gilson's (2011, 323) suggestion that an ambivalent conception of vulnerability allows us to grapple with the "repercussions of closing oneself to certain kinds of relations and situations." Most especially, I want to consider the implications of academics' efforts to close off relationships with those who fund our research.

Headaches in Qingshan

Part of the appeal of the Qingshan project to the IFF team that worked on it was the relief from countervailing imperatives of speed, scale, and value for money. At the outset of IFF's involvement in the project, the usual value for money and M&E imperatives were entirely absent. As the project progressed, however, IFF's apparent freedom from the usual logics of international development seemed to dissipate. Several months after running our first training workshop, the IFF team arranged what was to be a third workshop directly with a Qingshan community leader. Having made these arrangements, however, we received a communication from Arboreus explaining that all IFF activities in Qingshan must be preapproved by both Shehui and Arboreus. The workshop did eventually go ahead, but following the workshop, wrangling continued among Shehui, Arboreus, and IFF. In the weeks and months that followed, IFF staff became increasingly frustrated with project partners who we saw as bringing only headaches to the project.

Some IFF colleagues attributed the emerging difficulties of working in Qingshan to the incompetence of both Shehui and Arboreus. In one instance, Shehui's project manager Xuejian explained to the IFF team that Arboreus's training in Qingshan had attempted to impress on local farmers the importance of three-meter spacing between trees—a proposed guideline, Xuejian told us, that took no account of tree species or location. For IFF scientists, this absurdly global approach to tree propagation confirmed Arboreus's ignorance. The failure of the collaboration, however, can also be attributed to an incongruity between the approaches of IFF and its project partners. In addition to its incompetence, Arboreus's tree-spacing guidance evinced its professionalism: It reflected its understanding of environmental sustainability as a global, standardizable, and scalable process. Embracing the professional logics described in the preceding chapters, Arboreus is constantly contemplating how its activities might be replicated at ever-greater scales. Indeed, this is part of the reason that it wanted not only prior notification of IFF's activities but also written reports after each visit to Qingshan. On one occasion, impressed by a soil management booklet Susanna had produced specifically for Qingshan farmers, Arboreus asked whether IFF might create a similar set of booklets for it to use globally.

IFF, by contrast, approached the project much more narrowly: as a project in Qingshan. We were seldom concerned with the question of how to replicate or upscale activities in Qingshan. Qingshan workshops were each designed from scratch and as unique events. They did not adhere to a previously established protocol, nor did IFF staff imagine their workshops to be replicable. This contrasts with other projects where the imperatives of professional research and development led to protocols and tools designed for use anywhere and by anyone. Part of what initially excited IFF colleagues about Qingshan Agroforestry was that it seemed to present an opportunity to escape precisely these imperatives, to provide a space free from the expansive scale and value-for-money obsession of hyperprofessionalized research for development.

Shortly following the shambles of IFF's negotiation for permission to run our third workshop, Arboreus commissioned an audit of the project. One outcome of the audit was for the project sponsor, Metelli, to belatedly follow up on a suggestion that IFF had made several months earlier to establish experimental agroforestry demonstration plots in Qingshan. This is a proposal that Susanna and Matt had initially discussed with Mr. Wang, a local community leader. The idea was to lease a small plot of land from Mr. Wang and use it to plant a complex agroforestry system designed in collaboration with local farmers. Matt had shared this proposal with Metelli and Arboreus, but having heard no response, he assumed that the necessary financial support would not be forthcoming. In response to Metelli's belated interest, Matt and Susanna wrote up a detailed outline of their proposal, but some at IFF half-hoped Metelli would decline to fund the proposal and thereby give IFF an excuse to leave the project. The imperative to formalize project plans and project goals, as well as to undergo audits and manage budgets, returned the project squarely to the very space we thought we had escaped. IFF colleagues' enthusiasm for the project faded as it began to look less and less like an opportunity to engage farmers in Qingshan and more and more like a headache-inducing struggle of negotiating collaboration with the project's numerous supralocal partners.

The Qingshan project generated a particular kind of responsibility: significant energy was invested in crafting relationships with local Qingshan farmers. Personal commitments to farmers were central to IFF staff's dedication to the project as well as to scientists' disappointment in the project's ultimate collapse. This focus on relationships with farmers, however, eclipsed a separate relationship of accountability and responsibility that the project team actively wished to evade: the relationship to those who paid for the project. In her ethnography of a Japanese NGO in Myanmar, Chika Watanabe (pers. comm.) observes that in rejecting the status of "professional," NGO workers would evade the questions of accountability and responsibility to which professional work is tied. Similarly, when IFF staff embraced the Qingshan project as something that stands beyond the deliverables and reporting procedures that characterize much of IFF's work, we evaded questions of accountability and responsibility to anyone beyond Qingshan. We evaded the obligations and entanglements that professional relationships—relationships that entail being paid for work—ordinarily impose on scientific practice.

Opening Up and Closing Off Relationships

To be clear, I do not deny the many good reasons to escape entanglements not only with partners in the Qingshan project but with the broader professional imperatives variously exemplified by Alistair's marketing strategy and research for development tools and the Food Security Fund's MELA framework. Indeed, the invaluable relationships that IFF establish with farmers are in some cases undermined by the impersonal bureaucratic demands of funding organizations. When I returned to China in 2017 to conduct follow-up research, I discovered that the Agroforestry for Myanmar project had been terminated a year and a half early. Irreconcilable differences between IFF and its principal local partner had made management of the experimental agroforestry plots impossible. In meetings to plan the project's early termination, Matt requested that the Food Security Fund allow his team to continue with a mushroom training workshop that IFF had already promised to local farmers at a project site. For Matt's team, this promise had generated a personal obligation to the farmers in question, and this was an obligation that they wished to fulfill. The Food Security Fund's response, however, was that this workshop would not be an "effective use of funds."

This response can be understood in the logic of a Theory of Change and in the context of the Food Security Fund's agenda for impacts that are value for money. Initially the mushroom workshops were located—alongside IFF's other activities—in an integrated project Theory of Change that linked them to nationwide improvements in upland land management. Since the wider project had been disbanded, however, the mushroom workshops now stood in isolation. Shorn of their strategic relationship to the broader Theory of Change and to a concrete plan for upscaling, the mushroom workshops lost their claims to value for money. Unsurprisingly, the Food Security Fund saw little prospect of economically measurable return on investment from two international scientists running isolated workshops for twenty or so farmers. The logic of value for money superseded any sense of personal obligation to fulfill promises made to farmers who had been participating in the project.

There is an obvious value to evading the oversight of funders and institutions that fail to appreciate that which exceeds economistic evaluation. Enthusiasm for escaping these professional relationships, however, is not universal at IFF. As described in chapter 2, Jianming, a senior member of the corporate services team, departed IFF due to his loss of faith in the value of IFF-China's work. When he described IFF's lack of value to me, Jianming did so in a manner that echoed contemporary trends in M&E: he lamented IFF's failure to quantifiably demonstrate a return on investment for any of its projects. In addition to being a waste of his own energy, Jianming described IFF's work as a waste of taxpayers' money—the source, as he pointed out, of much of IFF's domestic and international research funding. Unlike the Qingshan project, which only sought to establish responsibility toward local project participants, Jianming pointed toward professional obligations to those who fund IFF's work.

Discussing his work at IFF over a coffee, another IFF staff member, Wasim, expressed a similar desire for IFF to take M&E more seriously. He described his frustrations with IFF colleagues evading obligations to the recently completed Asian Watersheds project—an initiative for which he had been project manager. Asian Watersheds involved multinational collaboration across IFF and two development organizations based in Central and South Asia. Unusually for an IFF-China colleague, Wasim lamented the lack of M&E in this project. He described how the project involved running a household survey across four sites with each project partner taking responsibility for at least one site. Rather than a new survey, however, a senior member of the IFF team decided that for the project's Chinese site, they would reuse data from an already completed household survey. To ensure uniformity across the other sites, Wasim explained, the other project partners decided to implement the same survey instrument from IFF's earlier household survey. As the project partners implemented this survey, however, it became apparent that the survey instrument was poorly designed and produced data of very little value to the project. It was a shame, in Wasim's view, that expediency took precedence over quality and that the lack of M&E associated with the project's funding allowed the IFF staff involved in this project to pass off work already completed as if it were contributing to the Asian Watersheds project.

Wasim regretted the donor's failure to impose stricter monitoring and evaluation. He told me that despite IFF appointing him as project manager, the lack of strictly enforced donor requirements left him impotent to force IFF colleagues and their project partners to effectively fulfill the goals originally envisaged for the project. Instead, as Wasim sees it, IFF only superficially satisfied what the donors had funded them to deliver while using the funds provided to support IFF's independent research agenda. This is an instance, in other words, where other IFF colleagues viewed donor requirements as nothing more than a headache. Their goal as such was to navigate the donor's reporting procedures in such a way that donor demands had minimal impact on the research that IFF scientists already had it in mind to pursue. Wasim's frustration was that paying heed to this donor's aspirations—rather than simply going it alone—might have enhanced IFF's work.

IFF's ability to spend the Asian Watersheds funding with relative liberty is another iteration of IFF scientists' desires to escape the constraints of oppressive professionalization. As Jianming reminds us, however, these constraints emerge out of IFF's economic relationship to those who fund its research. This is not to suggest that as scientists we should simply accept the demands and imperatives demanded by contemporary donors. The obsession with speed, quantification, and value for money typified by contemporary use of return on investment as the ultimate measure of efficacy is something that we must challenge. But to challenge impact agendas and evaluative rationales is not quite the same as to escape relationships with our financial benefactors. To escape the obligations tied to funding while simultaneously accepting funding is akin to seeking a free gift. In this respect, the attempt to evade the obligations of funding organizations mirrors the trust-evading practices of Chinese bureaucrats that, as described in chapter 3, so frustrate Tao and her IFF colleagues. Much as these bureaucrats seek to minimize the entanglements and vulnerabilities that a transaction might extend into the future, when scientists treat donor evaluations as a headache, they embrace a fantasy that the transaction of funds can be conducted in a manner that generates no entailments on or interruptions to the future paths of their research.

As Anna Tsing highlights, our energies as scholars and as scientists must exceed the narrow constraints of a professionalization typified by quantifiable certainties and the narrow, cold logic of value for money. My attention to the spaces that scientists have crafted in the space beyond these logics might suggest an optimistic take on this challenge: activities that attempt to exceed these limited rubrics and logics already abound. Though initiatives like Qingshan Agroforestry and proposals likes Jiaolong's beautiful Lengshan too often suffocate in the demands and expectations of contemporary institutions, these examples demonstrate that anthropologists are not alone in attempting to imagine and craft spaces that exceed the oppressive constraints of our professionalization. These are practices, moreover, that speak to anthropological and STS efforts to challenge the invulnerability of scientific authority.

There are, nevertheless, many grounds for ambivalence toward these spaces and toward the Qingshan project in particular. For all the IFF project team's anxiety-driven desire to create an environment in which local farmers would shape our activities in Qingshan, we were accountable only to ourselves in evaluating the success of this endeavor. This contrasts with our usual research practices, where we would expect to be accountable to our scientific peers as well. Our passion for this project was, moreover, heightened by the promise of evading the usual obligations to those funding our work. Spaces that transgress the boundaries of one's professional expertise can make the monopoly of scientific knowledge vulnerable and create valuable opportunities for interruption. But, to some extent at least, the Qingshan project also reflected a desire to work autonomously from those beyond our project team.

With research budgets closer to home under as much under pressure and scrutiny as international development funding, heightened desires for autonomy are perhaps as familiar to anthropologists as they are to agroforestry scientists. To give one example, when a colleague in North America recently received an award from a taxpayer-funded research grant, the anthropologist administering the award asked him to rewrite the project abstract with a right-wing legislator in mind. By rewriting the project abstract in this way, these two anthropologists hoped to avoid the project being made an example of by right-wing politicians trying to paint anthropological research as a waste of public resources. They created, in short, a facade that made it look like the research in question aligned with the values of legislators who might otherwise challenge their work. This rewriting of the abstract has much in common with the box ticking and headache navigating that I encountered at IFF. Much like some of IFF's actions described in this chapter, these two anthropologists sought to eliminate a project's potential vulnerability to a funder's interruptions. Unlike Qingshan Agroforestry, however, these anthropologists anticipated no parallel spaces of vulnerability to nonacademic communities.

There are of course very good reasons for undermining the power that state and corporate actors wield over research institutes and universities through the funds they control. I am not, in this respect, advocating that we simply accept the disproportionate power to interrupt wielded by philanthropic institutions like the Gates Foundation or by corrupt and compassionless governments like the one we have in the United Kingdom. Discomfort with the unjust and unequal power that such funders exercise, however, does not negate the fact that the wealth that supports our research is generated by people outside the academy—be they taxpayers or sweatshop workers producing Microsoft components. In this light, we might ask what it would mean to uncouple the refusal of toxic agendas for efficiency and return on investment from the fantasy of research invulnerable to the publics who pay for it—from the fantasy of funding as a free gift. Such a decoupling might allow us to start imagining and crafting a new foundation for our professional relationships with the publics on whom we are inevitably dependent.

A THEORY OF CHANGE
FOR THE SCIENCES

If I were to get out the note cards and pinboard Lesley used for her workshop and map out a Theory of Change for the sciences, I would begin with a vision statement that goes something like this: "Universities and the sciences help bring about a more just and more livable world." In contrast to the precise language of an M&E framework, this is a somewhat vague vision for the future. It has, nevertheless, been my guidepost and motivation for the questions that this book explores: questions about the horizons of scientific practice, about the agency we as scientists aspire to exert in the world, about what it could and should mean for people to trust in science and scientists, and about how we as scientists create opportunities to be shaped by the people and the ecosystems that surround us.

One of the exercises that Lesley asked colleagues to complete during her two-day workshop involved identifying potential barriers to reaching the vision atop IFF-China's Theory of Change. For my own Theory of Change, we could fill countless note cards with barriers not only to realizing a more just and livable world but also to universities and the sciences playing a positive role in the pursuit of one. At the time I was returning home from fieldwork in 2016, many of us were preoccupied with one such barrier: the threat from populist and post-truth politics exemplified by the recent Brexit referendum and US presidential election. The American Anthropological Association, for instance, invited submissions to its 2017 annual meeting on the theme "science under attack"; Bruno Latour declared a "second science war" (Vrieze 2017); and scientists mobilized to "march for science" (Slawson 2017). Though 2016 sometimes felt like it brought new and unprecedented challenges for the sciences, the narrative of an academy under siege and rallying calls to defend it were also deeply familiar. For at least two decades, academics had already been lamenting the slow death of the university at the hands of culture wars, corporatization, neoliberalism, and audit cultures (Ginsberg 2011; Newfield 2008; Readings 1996; Shore and Wright 1999; Thornton 2014).

Diagnoses and remedies to the challenges universities and the sciences face are diverse.[1] Many, however, focus on the space for intellectual freedom and creativity that we have lost at the hands of corporate-driven funding regimes and ever more invasive audit regimes. In this familiar narrative, the "free-thinking intellectual" of a bygone age has died at the hands of academic professionalization and "an avalanche of bureaucracy" (Grimshaw and Hart 1994, 237, 258). The creative freedom to "mess around" that universities once provided has been strangled by "[m]anagement [that] serve[s] the market rather than the craft and its practitioners" (Newfield 2003, 126, 219). In a similar spirit, anthropologists have called for us to "reclaim the professional autonomy and trust that audit practices appear to strip out of the workplace" (Shore and Wright 2015, 422). These laments and calls to action resonate with conversations I have observed and participated in with academics in universities across the United States, the United Kingdom, and elsewhere. They reflect widely shared anxieties and frustrations with the perceived demise of our vocations and of the modern university. These narratives have much in common, moreover, with the discontent of IFF colleagues who lament increasingly pervasive M&E regimes, ever-diminishing research budgets, and ever-receding time for doing research. As appealing as the idea of being left alone to get on with our work may be, however, this book's argument—my Theory of Change—is that remaking sciences and universities will require us to proliferate opportunities for interruption and, with that, vulnerability.

Embracing vulnerability is doubtless easier said than done in a context where many of our colleagues face extreme insecurity. In the United Kingdom, Research Excellence Framework (REF) evaluations and cuts in public funding combine to threaten not only research funding, but also livelihoods: Academics worry that if they do not satisfy peculiar criteria for "excellence," their jobs might be at risk. In some cases, entire programs and departments face closure for failing to meet dubious measures of success. More broadly, an increasing proportion of faculty across the world are employed on precarious, short-term contracts and hold little hope for a secure livelihood within the academy. On top of this, there is a huge gulf in job security and voice separating faculty from the student workers and nonfaculty staff who are a crucial but chronically undervalued part of the university (Potter 2022; Skallerup Bessette 2020a, 2020b; SOLR 2022). As acutely, IFF colleagues navigated the MELA framework against the background of an office struggling to fund staff salaries. The Food Security Fund imposing this new evaluation regime coincided with a moment of broader financial difficulties for IFF-China. Another of IFF's major projects had collapsed, several large funding applications had been unsuccessful, and IFF's international headquarters (which had significant difficulties of its own) was cutting funding to its regional offices. Some scientists had government-funded fellowships that guaranteed them a modest stipend, but many colleagues were dependent entirely on IFF's internal funds. By the time I had concluded my fieldwork in 2016, corporate services staff who were previously content in their jobs had begun discussing fears of redundancy and plans for alternative careers. As IFF's director explained to Matt, "Agroforestry for Myanmar is not a project that IFF can afford to fail." In this context, IFF colleagues have—like the administrators and bureaucrats described in chapter 3—developed a distinct aversion to vulnerability and an equally powerful desire to shore up some autonomy.

Far from opportunities for productive interruption to the project's trajectory, interactions with the Food Security Fund were driven—for reasons that are easy to sympathize with—by the imperative to keep their intrusions at bay. Through the MELA framework, the Food Security Fund had reoriented IFF's research toward a peculiar set of overdetermined, future-perfect goals. And with its protracted process for approving IFF's initial version of the framework, it had undermined IFF staff's confidence in assertions that MELA was "flexible and open to revision." With all of this against the gloomy prospect of what the project's collapse would mean for IFF-China and its ability to sustain its staff, it is unsurprising that Matt chose to hide his project's many difficulties from the Food Security Fund's gaze.

The most significant of these carefully obscured difficulties was IFF's relationship with the project's key local partner in Myanmar—Shada Forestry Institute. In his original grant proposal to the Food Security Fund, Jacob—who had preceded Matt as project lead—made much of the Shada Forestry Institute's role in the project. A major potential weakness of IFF's proposal, Jacob told me, was that IFF did not have a permanent staff in Myanmar. Having an in-country partner such as Shada Forestry Institute was therefore central to the project's viability. Moreover, Jacob explained, Shada Forestry Institute had close connections with local government, and this was a key selling point for a project that aspired to influence public policy. Two years into the project, however, IFF had made little progress toward signing a formal memorandum of understanding with Shada Forestry Institute. Communication had, moreover, increasingly broken down between Matt, who had by that time taken over as IFF's project manager, and his counterpart at Shada Forestry Institute. Matt decided to hide this problem from the Food Security Fund while he searched in vain for a solution.

Various ideas were mooted: finding personnel elsewhere in Shada Forestry Institute with whom collaboration might be easier to achieve, approaching other local institutions that might take Shada Forestry Institute's place, establishing IFF's own office in Myanmar and going it alone. When, more than three years into the project, Matt did finally reveal the full extent of the project's difficulties to the Food Security Fund, it was, as expected, quick to terminate the project.

The significance of this incident is not merely that Matt was right to fear the project being shut down or that the Food Security Fund was wrong to shut it down—by that time even Matt questioned whether the project should continue. Also significant is that this fear prevented Matt and his colleagues from openly communicating with the Food Security Fund. This lack of communication did not simply delay the project's collapse—which had not always been inevitable—it deprived IFF of potential support from the Food Security Fund. As a well-established and well-connected actor in Myanmar, had Matt been able to trust them to do so, the Food Security Fund might have assisted IFF in recruiting a new project partner or revising project plans before it was too late. Unable to trust the Food Security Fund to respond constructively and provide scope for IFF to revise its plans, however, Matt and his colleagues had little choice but to sustain the pretense that everything was going to plan. Just as Tao observes in relation to Chinese state bureaucrats, it is difficult to trust from a position of insecurity. When funding bureaucracies impose such intense insecurity on scientists, it is unsurprising that we respond by coveting independence and autonomy. It is in this respect that this example is most significant: it illuminates the misplaced desires for autonomy that bureaucratic headaches generate. Indeed, it is perhaps in fixating us on fantasies of autonomy—rather than in the deprivation of autonomy itself—that bureaucracies do most harm to the scientific work that they pretend to enhance.

Understood in relation to the Theory of Change I have set out in this book, autonomy is of course not the outcome to which bureaucracy is a barrier. In this respect, we might contrast Matt's despair with Agroforestry for Myanmar with the fruitful aspects of Qingshan Agroforestry. In the latter instance, Susanna made plans for soil management in Qingshan that were willfully vulnerable to the interruptions of local farmers. Unlike in the Agroforestry for Myanmar project, work in Qingshan was not tied to predetermined evaluation metrics. And IFF's ability to fund staff salaries was not, as it seemed to be in Agroforestry for Myanmar, tied to the successful realization of an overdetermined project plan. This gave Susanna and her colleagues a certain freedom from conventional donor-imposed demands such as value for money and scalability. But escape from these demands was significant not because it allowed Susanna and her colleagues to autonomously conduct whatever research they saw fit. Rather, it mattered because it allowed IFF staff to entangle themselves in relationships with rural communities—relationships that entailed ceding a significant degree of control over the project. Insofar as this project represented some of the potentials for what science might be outside the constraints of contemporary professionalization and bureaucratization, its distinctive characteristic was not the autonomy that Susanna enjoyed but the forms of vulnerability she generated. After all, scientific work at IFF depends not on pathways autonomously forged but on pathways that are shaped—via the interruptions of forums from peer review and PhD defenses to participatory workshops and walks in the forest—by the relationships in which scientists and their research are entangled and to which they are necessarily, and often willfully, vulnerable. In short, what audit cultures and our responses to them deprive us of is the ability to cultivate and to embrace forms of interruption and vulnerability that are—or at least could become—vital, creative, and potentially democratizing components of scientific practice.

In the language of Theory of Change, we might say that scientists' enjoying opportunities for interruption and vulnerability is my target outcome. Following through on the logic of Theory of Change, however, would mean turning this problematization into a pathway stretching from desired outcomes back to concrete actions. Thinking of barriers like audit cultures in relation to this outcome reminds us that dismantling destructive bureaucracies is a meaningful output only insofar as doing so serves as a stepping-stone toward building new, more adequate institutions and practices. Focusing on this outcome also demands answers to the question: "What activities and outputs would allow us to foster and embrace vulnerability and interruption?" In the spirit of the impressionistic thinking that Lesley encouraged colleagues to indulge during her Theory of Change workshop, here are a few tentative answers.

A first answer would be to transform audit and evaluation regimes. Institutions like the Food Security Fund that engender precarity and trust evasion clearly serve no one. Indeed, the ways in which such funders conventionally interrupt researchers are rather meaningless. When I have submitted research reports to organizations funding the ethnographic research for this book, for instance, the most I have ever received in response is a polite acknowledgment of receipt. This is a far cry from the unpredictable productivity of interruptions IFF colleagues experience in many of their interactions with colleagues and land-users. The only surprise one is likely to anticipate (or, rather, fear) from a funder is the one Matt ended up receiving from the Food Security Fund: notice that funding has been terminated. These are bureaucracies that interrupt in destructive but never productive ways. There is nothing inevitable about this. It is quite possible to imagine systems of accountability that provide meaningful opportunities for interruption. Indeed, one such system is peer review. This is a system that, when it works properly, delivers much more than a decision on whether a project lives or dies (though this is often at stake too): it provides feedback that shapes our work in important and unexpected ways. Of course, a problem with peer review is that it is generally only academic peers to whom we make our research vulnerable. But, as CLEAR's community peer-review system demonstrates, it is quite possible to develop analogous systems that engage communities beyond the academy (Liboiron, Zahara, and Schoot 2018). Community peer review is, in this respect, one model for what a meaningful academic bureaucracy could look like.

Transforming funding and evaluation regimes is no small undertaking. Can we really expect positive reforms in NSF or REF with the current crop of politicians overseeing them? If we cannot transform such institutions, we can nevertheless develop opportunities for interruption and vulnerability elsewhere. In this respect, so long as funder bureaucracies provide only meaningless interruption, the kind of evasion Matt pursued in relation to the Food Security Fund is perhaps not only understandable but advisable too. Navigating autonomy from meaningless evaluation regimes may be a precondition to cultivating meaningful forms of vulnerability elsewhere. Again, community peer review is an example of what this might look like. One could imagine a world in which such practices were a core component of audit regimes like REF, but equally one could (and CLEAR has) develop such a practice despite powerful institutions that expect accountability to happen elsewhere and otherwise. Indeed, many of the examples I have described of productive vulnerability and interruption at IFF have emerged in spaces alongside or beyond the usual strictures of academic and development institutions.

In this regard, another answer to the question of activities would be to follow through on and proliferate some of the ideas and practices that I encountered at IFF. We might, for instance, seek to replicate the ways in which Susanna made her plans and her knowledge vulnerable to the interruptions of local farmers in Qingshan. We might also take Jiaolong's lead and develop new ways to communicate our "research results to our primary and middle school classmates, to our classmates' classmates, to our kin's kin." As I described in chapter 1, Jiaolong's call to action here was inspired by her encounter with a scholar who responded to her having read his PhD dissertation by asking: "Why on earth would you want to read my dissertation?" One interpretation of this response is that it expressed humility—that he was saying, "My research is not so great. You'd be better off reading someone else's." There is perhaps a dimension of this to that interaction. It might, however, be read simultaneously as a refusal to vulnerability. To joke that one's research is not worth reading is an easy way to end a conversation that one fears could end in disapproval, even ridicule. There is something quite familiar about this response. Certainly, I have found myself chatting with the old classmates Jiaolong urges us to engage and done something similar: I have deflected questions about what I do by intimating that my research really is not that interesting. On other occasions, I have engaged those same classmates and discovered unexpected resonances between what I study and their own professional worlds. It turns out, for instance, that when it comes to frustrations with evaluation regimes, primary school teachers in the United Kingdom have much in common with agroforestry scientists in China. There are obvious limitations to practices that engage only with the nonacademics we know from high school. Indeed, Jiaolong invokes our classmates as part of a much wider public, one that goes beyond our immediate social group. Nevertheless, one way for us to build opportunities for vulnerability and interruption would be to make more of the conversations many of us already have with nonacademics. Whether such ad hoc conversations be over dinners with old school friends or with strangers who strike up conversations on the bus, these are everyday opportunities for interruption that could enrich our work.

Jiaolong's orientation to public engagement is, of course, one that has its parallels elsewhere.[2] Her embrace of the personal and of storytelling as part of—rather than alien to—scientific practice has an affinity with feminist STS (Haraway 1991; Hubbard 1988; Keller 1985; Roy 2018; Subramaniam and Willey 2017a). Likewise, her proposal for film and art as mediums for public engagement in the Eco-Friendly Rubber project brings to mind the diverse ways in which scientists, STS scholars, and anthropologists are already using these mediums to craft new forms of collaboration (Collins, Durington, and Gill 2017; da Costa and Philip 2008; Hegel and Cantarella 2021; Hong 2021; Smith et al. 2021).[3] The cases for community peer review, feminist sciences, and artistic collaborations have been made on numerous grounds and serve vital outcomes beyond the scope of this book. Understood in relation to this book's Theory of Change, my point is simply to additionally appreciate such practices in relation to the struggle against the suffocating and counterproductive bureaucracy of contemporary research institutions. Most ambitiously, this appreciation might mean advocating these existing practices as models for transformed bureaucracies—bureaucracies that make us vulnerable to meaningful and potentially productive interruptions from wider publics. More modestly, appreciating these practices in relation to audit cultures might help us disentangle the goal of invulnerability to meaningless impact evaluation frameworks from misplaced desires for invulnerability and autonomy in general. Along with some of the activities I have described at IFF, these initiatives serve as a reminder that it is difficult but nevertheless still possible and vital to sustain opportunities for vulnerability and interruption in the shadow of headache-inducing bureaucracies.

Notes

INTRODUCTION

1. Indeed, my own analysis of Chinese bureaucracy is inspired by Matthew Hull's (2012) ANT-inflected ethnography of Pakistani bureaucracy.

2. A similar attention to temporality has long been a feature of the anthropology of gifts. Though their ethnographies of exchange differ in many ways, Nancy Munn's (1986, 58) analysis of the "complex interplay of the incommensurable spacetimes of different [Gawan] transactions," and Marilyn Strathern's (1988, 305) analysis of agency in the "temporalized transactions" of Melanesian exchange are both acutely sensitive to the actions and relationships that their interlocutors make visible as the cause of any given transaction, as well as to the effects that these transactions anticipate. In this regard, Munn and Strathern exemplify (and have inspired) a wider anthropological sensitivity to the ways exchange both is structured by and gives effect to temporal forms (e.g., Chu 2010; Keane 1997; Miyazaki 2004; Watanabe 2015; Weiner 1976).

3. Michelle Murphy (2021) draws inspiration for working "with and against technoscience" from Frantz Fanon, a figure she suggests we should place at the heart of STS's genealogy.

4. For the sake of consistency, I use the single English term "headache" except in those cases where I am directly quoting IFF staff. Though this may seem like a strange translation for China scholars, it is the one used by my bilingual interlocutors at IFF. Another approach would have been to use the untranslated term *mafan*—as many China anthropologists do for terms like *guanxi*—but the English translation adopted by IFF colleagues has the benefit of familiarity to readers who do not speak Chinese.

5. As I explore in chapter 2, there are times when things like peer review and PhD defenses that are imagined as core scientific activities also become subsumed within this logic of a bureaucratic headache. Conversely, as I describe in chapter 5, for proponents of Theory of Change like Lesley and Bob, Theory of Change is not a headache or bureaucracy; it is a tool that scientists might adopt as a vital component of their research practice. In this book, I follow Matt, Ruyue, and like-minded colleagues in categorizing Theory of Change with bureaucracy. At the same time, however, I hope to trouble the assumption (shared by many colleagues in universities elsewhere in the world) that bureaucracy is inherently that different from or detrimental to scientific practice.

6. Andrew Mathews (2011, 13) has critiqued anthropologists of development for adopting similar lines of inquiry because they "too easily accept modernist bureaucracies' rhetoric of general or abstract knowledge, even as they criticize them for failing to live up to their proclaimed projects." Mathews calls for us to pay greater attention instead to "how making knowledge and ignorance are partially intentional practices." In the context of my ethnography, the limitation to approaching bureaucracy as a project of general or abstract knowledge creation is not that it ignores the intentionality or self-awareness of ignorance creation (also see Welker 2014, 183–213). Rather, it is the premise that we should think of bureaucratic knowledge in terms of adequation to reality (cf. Hetherington 2011; Maurer 2005).

7. Another rich body of examples might be found in the diverse and fast-growing field of collaborative anthropology (Boyer and Marcus 2021; de la Cadena 2015; Faubion and

Marcus 2009; Lassiter 2005; Rappaport 2005). As I highlight in this book's conclusion, a proposal of Jiaolong's to bringing film and artwork into Eco-Friendly Rubber resonates with the collaborative potential some anthropologists identify in artistic and visual methods and mediums (Collins, Durington, and Gill 2017; Hegel and Cantarella 2021; Hong 2021; Smith et al. 2021). While my own disciplinary bias is toward anthropology and STS, there are surely examples to be found elsewhere too.

8. My starting salary was 4,000RMB (approximately US$600) per month, rising to 6,000RMB (US$900) in my second year.

9. Despite stereotypes to the contrary, the analysis I elaborate in chapters 6 and 7 of this book highlights that this kind of entanglement is often just as invaluable to natural science research.

10. This has far from entirely redressed the unevenness of how accessible and meaningful drafts of this work have been to a group of colleagues with very different levels of familiarity and comfort with English, with academic writing, and with anthropology in particular. My own research practices may, in this respect, have unintentionally reproduced academic hierarchies that I have elsewhere critiqued for devaluing the intellectual contributions of nonfaculty staff (McLellan 2021b).

1. THEORY OF CHANGE

1. Who is targeted as a next-user will depend in large part on who one identifies as the actors most capable of or most necessary to affecting target impacts. The next-users we identified during Lesley's workshop included national and local governments, the media, research organizations, NGOs, intergovernmental organizations (IGOs), and development donors. The idea that a donor is a next-user recognizes the role that donors play in funding IFF's, as well as in funding the work of other next-users that IFF targets within its Theory of Change. Identifying donors as next-users recognizes the power of financial resources over research and development work.

Conversely, IFF might itself stand as a next-user in the eyes of other institutions. As I will describe in the chapter, the Food Security Fund demanded that IFF follow a burdensome M&E framework that included a project-level Theory of Change. This imposition could be understood within the logic of the Food Security Fund's own institutional level Theory of Change and impact pathway. Based on the notion that M&E contributes to development impacts, the Food Security Fund's strategy targets local capacity in M&E as a target outcome. Put in the terms used in Lesley's outcomes thinking workshop, IFF learning and adopting new M&E methods is a next-user change in knowledge, attitude, and skills that will help the Food Security Fund on its pathway to achieving its vision for improved livelihoods in Myanmar.

2. In this context, the STS idea that a scientist's success lies in the ability to interest a multitude of actors (Callon 1984; Latour 1987) is transformed from path-breaking analysis into a program for action. That is to say, the multistakeholder collaborative dimension to outcomes thinking turns the STS analytic of coproduction into an instrumental strategy (Jasanoff 2004). In this regard, multistakeholder platforms for rural China have a close affinity to the mutual learning processes adopted in French medical research. As Vololona Rabeharisoa and Michel Callon (2004, 158) put it, if "co-production implies a collective action and mobilization, then the work for organizing it imposes itself as a prominent issue." This is no accidental coalescence of ideas: Lesley herself pointed the workshop participants to a piece of STS scholarship arguing that scientists must "consciously support the coproduction of knowledge" (Cash, Borck, and Patt 2006, 465).

3. The open-endedness that Jacob describes is appropriate to the moment of planning research as well as to the process of implementing that plan. This stands in contrast to

the descriptive modality of a research paper that might attempt to stabilize or close a certain matter of fact. A good research proposal is necessarily open to revision in a way that is not necessarily the case for good research findings. These modalities coexist, however, because they are appropriate to different stages of a research project (Latour 1987; Pickering 1995, 113–56). There is a sense, therefore, in which the retrospective description of methods in the research paper eclipses the provisional nature of methods in the research proposal. This eclipsing, however, comes only after the research project has been completed: the provisionalism and closure are sequential to one another. By contrast, as Jacob's frustration with MELA highlights, the decisive retrospective closure of M&E frameworks imposes itself right from the start. By beginning with the MELA framework's advance determination of project goals as the basis for a future decisive retrospective, MELA erases the provisional status Jacob would have given the project plan.

4. We might also consider Alistair's brand-building strategy as a kind of instrumentalization of the discursive power that scholars such as Arturo Escobar (1995) and James Ferguson (1994) reveal within concepts such as "development." When Ferguson (1994) describes the "instrument effect" of development discourse, the proliferation of this discourse is something "without a master plan," something "unintended." Alistair's brand, by contrast, is very much built from a plan (Hilgartner 2015). Reworking Escobar (1995, 216), we might say that Alistair's design for agroforestry as a brand is driven by an implicit recognition that "changing the order of discourse is a political question that entails the collective practice of social actors." Brand building, in this respect, implies a conscious effort to activate the collective agency of agroforestry scientists, empowering them to shape the world.

2. BUREAUCRACY AS INTERRUPTION

1. In ordinary usage, *biaomei* means a cousin or a close friend. The character *biao* on its own, however, means "form," while *mei* means "younger sister." This colleague's nickname plays on this double meaning.

2. Here Jianming echoes a dominant logic of evaluation in contemporary research for development. I return to this logic—and Jianming's invocation of it—in chapters 5, 6, and 7.

3. Rottenburg here echoes a wider turn to Bruno Latour's ANT to make sense of audit and evaluation regimes (Davis, Kingsbury, and Merry 2012; Merry 2016; Mosse 2005).

4. Much of Rottenburg's critique focuses on the idea that evaluation creates "fictions"—a term Rottenburg uses in the sense of a falsehood rather than in the sense of a fabrication or an as if (cf. Geertz 1973; Leach 1965).

5. This is of course not to say that visual imagery is necessarily absent from how IFF's counterparts in government offices or development institutions imagine the same bureaucratic work. Indeed, transparency is a rhetorical component of China's ongoing anticorruption campaign. In a slightly different manner, visual imagery is also key to the Food Security Fund's conceptualization of Theory of Change. The first step in creating a Theory of Change is, after all, to imagine a "vision" of the world one wishes to bring into being. My concern in this chapter, however, is with how IFF staff conceptualize and act on bureaucratic processes.

6. Writing with Davydd Greenwood, Susan Wright elsewhere identifies the more fundamental challenge of transforming the ownership and organizational structure of the university—a challenge that extends far beyond the mere defense of the university against the encroachment of audit (Wright and Greenwood 2017).

7. See chapter 1.

8. Michael Mascarenhas (2017, 123) makes an analogous argument in the context of international humanitarianism, suggesting that "[i]f . . . recent rounds of reforms [to the World Bank's funding system] were intended to engender trust and more accountability between humanitarian workers and donors, then they . . . have failed."

9. There are, in fairness, a good number of grants and fellowships that do provide substantive—and often extremely valuable—feedback on rejected and accepted proposals. I have, however, encountered as many that do not do this, and I have yet to encounter a grant progress report that garners anything beyond a polite acknowledgment of receipt.

5. ACCELERATING, UPSCALING, DESKILLING

1. In this respect, decision analysis shares some similarities with "expert elicitation" methods employed in climate science (see O'Reilly 2017, chap. 3).

2. See chapter 1.

3. RCTs were briefly mooted as a method to assess the impact of IFF's Agroforestry for Myanmar project. The Food Security Fund, however, ultimately decided that IFF should focus on evaluating next-user outcomes rather than on the end-user impacts that RCTs are generally used to study. Lucas, an M&E specialist at IFF headquarters, also proposed that IFF-China use a version of RCT for as many of its projects as possible. Like many of his headquarters colleagues, Lucas traveled to IFF's China office to preach the virtues of the RCT and M&E activities for which he has global responsibility. Not least because of the significant cost involved in carrying out RCTs, no one at IFF-China acted on Lucas's proposal.

4. Ingold (2000, 315) further distinguishes what he calls "tools" from "technique" and "technology." I use "tool" and "technology" interchangeably to reflect the fact that what development professionals refer to as a tool would in Ingold's terms be a technology.

5. In contrast to Carlos's highly formal tool, James Fairhead and Melissa Leach's (1996; Fairhead and Scoones 2005) work in West Africa demonstrates how ethnographic methods might help agronomists and soil scientists develop reflexive and comparative approaches to understanding and engaging local knowledge of soils and ecosystems.

6. This echoes the gradual process through which cell biologists "acquir[e] the dexterity to notice" (Myers and Dumit 2011, 252; also Dumit 2018; Myers 2015). Like Barbara McClintock's "feeling for the organism" (Keller 1983), a cell biologist's (or a soil scientist's) "capacity to be affected [by her object of study] is acquired over time" (Myers and Dumit 2011, 252).

7. This situation is perhaps not unique to agricultural research and development. In the context of public health, medical anthropologist Carolyn Smith-Morris (2016, 199) similarly laments, "My fear is that qualitative research seems to have lost the battle for health care and federal-funding relevance."

8. David Mosse (2005, 178) makes a similar observation in relation to scale and value for money in the context of rural development projects in India: "'[V]alue-for-money . . . demanded (i) the transformation of a diffuse collection of isolated projects into a rational and integrated programme, (ii) expansion and replication of successful rural development models on a scale "capable of attracting the attention . . . of district, state and national politicians, administrators, technocrats and businessmen responsible . . ." and (iii) the use of these to influence Indian government policy. No project could be justified in and of itself. Projects like IBRFP had to expand, replicate and 'be capable of wider influence to justify their costs.'"

6. A FEEL FOR THE ENVIRONMENT

1. Another Nobel laureate in physics, Felix Hoenikker, offered a similar celebration of play when, during his Nobel Prize acceptance speech, he proclaimed, "I stand before you now because I never stopped dawdling like an eight-year-old on his way to school. Anything can make me stop and look and wonder, and sometimes learn" (Vonnegut 1963, 17).

2. Joseph Dumit (2021) uses this term to describe a "substance as methods" research strategy that is quite different to the IFF research methods to which I am applying the same label. What these methods share, however, is a grounding in playfulness and the opportunities generated by being surprised.

7. GENERATING AND EVADING VULNERABILITY

1. Here, IFF scientists' views resonate with and in fact often draw directly on historical and social scientific critiques of modernist, agriculture, forestry, and land management (for example, Scott 1998; Shapiro 2001; Sivaramakrishnan 1999; Tsing 2005).

2. Julie's point is perhaps similar to one made by Evelyn Fox Keller (1983, 201) in her biography of Barbara McClintock: "A deep reverence for nature, a capacity for union with that which is to be known—these reflect a different image of science from that of a purely rational enterprise. Yet the two images have coexisted throughout history. We are familiar that a form of mysticism—a commitment to the unity of experience, the oneness of nature, the fundamental mystery of the laws of nature—plays an essential role in the process of scientific discovery."

3. See chapter 1.

4. In this respect, while we stepped outside certain of the logics of research for development tools, there is also a sense in which we embodied another tendency of impact agendas similar to those that drive the proliferation of prefabricated, scalable tools. We were a "party of professionals rolling up [our] sleeves and getting down to problem-solving" (Strathern 2006, 195).

CONCLUSION

1. For some of the most inspiring recent writing on the future of the sciences and universities, see la paperson 2017; Liboiron 2021; Roy 2018; Skallerup Bessette 2019; Stengers 2018.

2. As I highlight in the introduction, the feminist, anthropological, and STS literatures that Jiaolong's ideas parallel were a key inspiration for my ethnographic attention to her work in the first place.

3. My references here have a clear disciplinary bias: I have cited examples that are primarily connected to anthropology and STS and that are reported in published literature. I suspect one could find as many examples elsewhere, but in the impressionistic spirit of Lesley's workshop, I have taken already familiar (to me) references as illustrative examples rather than attempt a systematic, exhaustive, or representative review. I suspect and hope that this book will bring to some readers' minds examples of many more existing initiatives and practices that are already creating spaces for vulnerability and interruption.

References

Adam, Barbara. 1998. "Values in the Cultural Timescapes of Science." *Journal for Cultural Research* 2 (2–3): 385–402. https://doi.org/10.1080/14797589809359306.

Adams, Vincanne. 2013. "Evidence-based Global Public Health: Subjects, profits, erasures." In *When People Come First: Critical Studies in Global health*, edited by João Guilherme Biehl and Adriana Petryna, 54–90. Princeton: Princeton University Press. http://dx.doi.org/10.23943/princeton/9780691157382.003.0004.

Adams, Vincanne, ed. 2016. *Metrics: What Counts in Global Health*. Durham: Duke University Press. https://doi.org/10.1215/9780822374480.

Adams, Vincanne, Nancy J. Burke, and Ian Whitmarsh. 2014. "Slow Research: Thoughts for a Movement in Global Health." *Medical Anthropology* 33 (3): 179–97. https://doi.org/10.1080/01459740.2013.858335.

Bemme, Dörte. 2019. "Finding 'What Works': Theory of Change, Contingent Universals, and Virtuous Failure in Global Mental Health." *Culture, Medicine, and Psychiatry* 43 (4): 574–95. https://doi.org/10.1007/s11013-019-09637-6.

Berg, Maggie, and Barbara K. Seeber. 2016. *The Slow Professor: Challenging the Culture of Speed in the Academy*. Toronto: University of Toronto Press. http://dx.doi.org/10.3138/9781442663091.

Biruk, Crystal. 2018. *Cooking Data: Culture and Politics in an African Research World*. Durham: Duke University Press. https://doi.org/10.1215/9780822371823.

Bornstein, Lisa. 2006. "Systems of Accountability, Webs of Deceit? Monitoring and Evaluation in South African NGOs." *Development* 49 (2): 52–61. https://doi.org/10.1057/palgrave.development.1100261.

Boyer, Dominic, and George E. Marcus, eds. 2021. *Collaborative Anthropology Today: A Collection of Exceptions*. Ithaca: Cornell University Press. https://doi.org/10.7591/cornell/9781501753343.001.0001.

Brenneis, Don. 1994. "Discourse and Discipline at the National Research Council: A Bureaucratic Bildungsroman." *Cultural Anthropology* 9 (1): 23–36. https://doi.org/10.1525/can.1994.9.1.02a00020.

Brenneis, Don. 2006. "Reforming Promise." In *Documents: Artifacts of Modern Knowledge*, edited by Annelise Riles, 41–70. Ann Arbor: University of Michigan Press.

Brown, Hannah, and Maia Green. 2017. "Demonstrating Development: Meetings as Management in Kenya's Health Sector." *Journal of the Royal Anthropological Institute* 23 (S1): 46–63. https://doi.org/10.1111/1467-9655.12593.

Butler, Judith. 2015. *Notes toward a Performative Theory of Assembly*. Cambridge, MA: Harvard University Press. http://dx.doi.org/10.4159/9780674495548.

Butler, Judith, Zeynep Gambetti, and Leticia Sabsay, eds. 2016. *Vulnerability in Resistance*. Durham: Duke University Press. https://doi.org/10.1215/9780822373490.

Callon, Michel. 1984. "Some Elements of a Sociology of Translation: Domestication of the Scallops and the Fishermen of St Brieuc Bay." *The Sociological Review* 32 (1): 196–233. https://doi.org/10.1111/j.1467-954X.1984.tb00113.x.

Cash, David W., Jonathan C. Borck, and Anthony G. Patt. 2006. "Countering the Loading-Dock Approach to Linking Science and Decision Making Comparative Analysis of El Niño/Southern Oscillation (ENSO) Forecasting Systems." *Science, Technology & Human Values* 31 (4): 465–94. https://doi.org/10.1177/0162243906287547.

Cavanaugh, Jillian R. 2016. "Documenting Subjects: Performativity and Audit Culture in Food Production in Northern Italy." *American Ethnologist* 43 (4): 691–703. https://doi.org/10.1111/amet.12384.

Chambers, Robert. 1983. *Rural Development: Putting the Last First*. London: Longman. https://doi.org/10.4324/9781315835815.

Cho, Isabella B., and Ariel H. Kim. 2022. "38 Harvard Faculty Sign Open Letter Questioning Results of Misconduct Investigations into Prof. John Comaroff." *The Harvard Crimson*, 4 February 2022. https://www.thecrimson.com/article/2022/2/4/comaroff-sanctions-open-letter/.

Chu, Julie Y. 2010. *Cosmologies of Credit: Transnational Mobility and the Politics of Destination in China*. Durham: Duke University Press. https://doi.org/10.1215/9780822393160.

Chua, Roy Y. J., Michael W. Morris, and Paul Ingram. 2009. "Guanxi vs Networking: Distinctive Configurations of Affect- and Cognition-Based Trust in the Networks of Chinese vs American Managers." *Journal of International Business Studies* 40 (3): 490–509. https://doi.org/10.1057/palgrave.jibs.8400422.

Chun, Allen. 2018. *Forget Chineseness: On the Geopolitics of Cultural Identification*. Albany: SUNY Press.

Collins, Harry, and Robert Evans. 2002. "The Third Wave of Science Studies: Studies of Expertise and Experience." *Social Studies of Science* 32 (2): 235–96. https://doi.org/10.2307/3183097.

Collins, Harry, Robert Evans, and Martin Weinel. 2017. "STS as Science or Politics?" *Social Studies of Science* 47 (4): 580–86. https://doi.org/10.1177/0306312717710131.

Collins, Samuel Gerald, Matthew Durington, and Harjant Gill. 2017. "Multimodality: An Invitation." *American Anthropologist* 119 (1): 142–46. https://doi.org/10.1111/aman.12826.

Collmann, Jeff. 1988. *Fringe-Dwellers and Welfare: The Aboriginal Response to Bureaucracy*. Brisbane: University of Queensland Press.

Cooke, Bill, and Uma Kothari, eds. 2001. *Participation: The New Tyranny?* London: Zed Books.

CSIRO. n.d. "Ensuring We Deliver Impact." Accessed August 6, 2018. https://www.csiro.au/en/About/Our-impact/Our-impact-model/Ensuring-we-deliver-impact.

da Costa, Beatriz, and Kavita Philip, eds. 2008. *Tactical Biopolitics: Art, Activism, and Technoscience*. Cambridge, MA: MIT Press. http://dx.doi.org/10.7551/mitpress/9780262042499.001.0001.

Davis, Kevin E., Benedict Kingsbury, and Sally Engle Merry. 2012. "Indicators as a Technology of Global Governance." *Law & Society Review* 46 (1): 71–104. https://doi.org/10.1111/j.1540-5893.2012.00473.x.

de la Cadena, Marisol. 2015. *Earth Beings: Ecologies of Practice across Andean Worlds*. Durham: Duke University Press. https://doi.org/10.1215/9780822375265.

de Laet, Marianne de, and Annemarie Mol. 2000. "The Zimbabwe Bush Pump: Mechanics of a Fluid Technology." *Social Studies of Science* 30 (2): 225–63. https://doi.org/10.1177/030631200030002002.

Dumit, Joseph. 2018. "Notes toward Critical Ethnographic Scores: Anthropology and Improvisation Training in a Breached World." In *Between Matter and Method: Encounters in Anthropology and Art*, edited by Gretchen Bakke and Marina Peterson, 51–72. New York: Bloomsbury Academic. http://dx.doi.org/10.4324/9781003084792-4.

Dumit, Joseph. 2021. "Substance as Method: Bromine, for Example." In *Reactivating Elements*, edited by Dimitris Papadopoulos, Maria Puig de la Bellacasa, and Natasha Myers, 84–107. Durham: Duke University Press. https://doi.org/10.1215/9781478021674-005.

Escobar, Arturo. 1995. *Encountering Development: The Making and Unmaking of the Third World*. Princeton: Princeton University Press. https://doi.org/10.1515/9781400839926.

Escobar, Arturo. 1999. "The Invention of Development." *Current History* 98: 382. https://doi.org/10.1525/curh.1999.98.631.382.

Fairhead, James, and Melissa Leach. 1996. *Misreading the African Landscape: Society and Ecology in a Forest-Savanna Mosaic*. Cambridge: Cambridge University Press. https://doi.org/10.1017/CBO9781139164023.

Fairhead, James, and Ian Scoones. 2005. "Local Knowledge and the Social Shaping of Soil Investments: Critical Perspectives on the Assessment of Soil Degradation in Africa." *Land Use Policy* 22: 33–41. https://doi.org/10.1016/j.landusepol.2003.08.004.

Fanon, Frantz. 1963. *The Wretched of the Earth*. Translated by Constance Farrington. New York: Grove Press.

Farmer, Paul. 2016. "The Second Life of Sickness: On Structural Violence and Cultural Humility." *Human Organization* 75 (4): 279–88. https://doi.org/10.17730/1938-3525-75.4.279.

Faubion, James D, and George E Marcus, eds. 2009. *Fieldwork Is Not What It Used to Be: Learning Anthropology's Method in a Time of Transition*. Ithaca: Cornell University Press. http://dx.doi.org/10.7591/9780801463594.

Fei, Xiaotong. 1992. *From the Soil: The Foundations of Chinese Society*. Translated by Gary G. Hamilton and Zheng Wang. Berkeley: University of California Press. http://dx.doi.org/10.1525/9780520912489.

Ferguson, James. 1994. *The Anti-Politics Machine: "Development," Depoliticization, and Bureaucratic Power in Lesotho*. Minneapolis: University of Minnesota Press.

Galvin, Shaila Seshia. 2018. "The Farming of Trust": *American Ethnologist* 45 (4): 495–507. https://doi.org/10.1111/amet.12704.

Gao, Bingzhong. 2014. "How Does Superstition Become Intangible Cultural Heritage in Postsocialist China?" *Positions* 22 (3): 551–72. https://doi.org/10.1215/10679847-2685377.

Geertz, Clifford. 1973. *The Interpretation of Cultures: Selected Essays*. New York: Basic Books.

Gieryn, Thomas F. 1983. "Boundary-Work and the Demarcation of Science from Non-Science: Strains and Interests in Professional Ideologies of Scientists." *American Sociological Review* 48 (6): 781–95. https://doi.org/10.2307/2095325.

Gilson, Erinn. 2011. "Vulnerability, Ignorance, and Oppression." *Hypatia* 26 (2): 308–32. https://doi.org/10.1111/j.1527-2001.2010.01158.x.

Ginsberg, Benjamin. 2011. *The Fall of the Faculty: The Rise of the All-Administrative University and Why It Matters*. Oxford: Oxford University Press. http://dx.doi.org/10.1093/oso/9780199782444.001.0001.

Giri, Ananta. 2000. "Audited Accountability and the Imperative of Responsibility: Beyond the Primacy of the Political." In *Audit Cultures: Anthropological Studies in Accountability, Ethics, and the Academy*, edited by Marilyn Strathern, 173–95. London: Routledge.

Grimshaw, Anna, and Keith Hart. 1994. "Anthropology and the Crisis of the Intellectuals." *Critique of Anthropology* 14 (3): 227–61. https://doi.org/10.1177/0308275X9401400301.

Grosz, Elizabeth. 1999. "Darwin and Feminism: Preliminary Investigations for a Possible Alliance." *Australian Feminist Studies* 14 (29): 31–45. https://doi.org/10.1080/08164649993317.

Grosz, Elizabeth. 2005. *Time Travels: Feminism, Nature, Power*. Durham: Duke University Press. https://doi.org/10.1215/9780822386551.

Gupta, Akhil. 2012. *Red Tape: Bureaucracy, Structural Violence, and Poverty in India*. Durham: Duke University Press. https://doi.org/10.1215/9780822394709.

Guyer, Jane I. 2007. "Prophecy and the near Future: Thoughts on Macroeconomic, Evangelical, and Punctuated Time." *American Ethnologist* 34 (3): 409–21. https://doi.org/10.1525/ae.2007.34.3.409.

Hall, Elizabeth F., and Todd Sanders. 2015. "Accountability and the Academy: Producing Knowledge about the Human Dimensions of Climate Change." *Journal of the Royal Anthropological Institute* 21 (2): 438–61. https://doi.org/10.1111/1467-9655.12162.

Haraway, Donna J. 1988. "Situated Knowledges: The Science Question in Feminism and the Privilege of Partial Perspective." *Feminist Studies* 14 (3): 575–99. https://doi.org/10.2307/3178066.

Haraway, Donna J. 1991. *Simians, Cyborgs, and Women: The Reinvention of Nature*. New York: Routledge. http://dx.doi.org/10.4324/9780203873106.

Haraway, Donna J. 1997. *Modest_Witness@Second_Millennium.FemaleMan_Meets_OncoMouse*. New York: Routledge.

Harding, Sandra G. 1991. *Whose Science? Whose Knowledge?: Thinking from Women's Lives*. Ithaca: Cornell University Press. https://doi.org/10.7591/9781501712951.

Harding, Sandra G. 2015. *Objectivity and Diversity: Another Logic of Scientific Research*. Chicago: University of Chicago Press. http://dx.doi.org/10.7208/chicago/9780226241531.001.0001.

Hathaway, Michael J. 2013. *Environmental Winds: Making the Global in Southwest China*. Berkeley: University of California Press. https://doi.org/10.1525/california/9780520276192.001.0001.

He, Xin. 2012. "Black Hole of Responsibility: The Adjudication Committee's Role in a Chinese Court." *Law & Society Review* 46 (4): 681–712. https://doi.org/10.1111/j.1540-5893.2012.00514.x.

Hegel, Christine, and Luke Cantarella. 2021. "Ethnographic Reentanglements in the Collaborative Ecologies of Film and Contact Improvisation." In *Collaborative Anthropology Today: A Collection of Exceptions*, edited by Boyer, Dominic, and George E. Marcus, 54–74. Cornell University Press. https://doi.org/10.7591/cornell/9781501753343.003.0004.

Hertz, Ellen. 1998. *The Trading Crowd: An Ethnography of the Shanghai Stock Market*. Cambridge: Cambridge University Press. http://dx.doi.org/10.1017/CBO9781139166850.

Herzfeld, Michael. 1992. *The Social Production of Indifference: Exploring the Symbolic Roots of Western Bureaucracy*. Chicago: University of Chicago Press. http://dx.doi.org/10.4324/9781003135029.

Hetherington, Kregg. 2005. *Guerrilla Auditors: The Politics of Transparency in Neoliberal Paraguay*. Durham: Duke University Press. https://doi.org/10.1215/9780822394266.

Hilgartner, Stephen. 2015. "Capturing the Imaginary: Vanguards, Visions, and the Synthetic Biology Revolution." In *Science & Democracy: Knowledge as Wealth and Power in the Biosciences and Beyond*, edited by Stephen Hilgartner, Clark Miller, and Rob Hagendijk, 33–55. New York: Routledge.

Hong, Emily. 2021. "The Multiply Produced Film: Collaboration, Ethnography, and Feminist Epistemology." *Cultural Anthropology* 36 (4): 649–80. https://doi.org/10.14506/ca36.4.09.

Hubbard, Douglas W. 2014. *How to Measure Anything: Finding the Value of Intangibles in Business.* 3rd ed. Hoboken: John Wiley and Sons.

Hubbard, Ruth. 1988. "Science, Facts, and Feminism." *Hypatia* 3 (1): 5–17. https://doi.org/10.1111/j.1527-2001.1988.tb00053.x.

Hull, Matthew S. 2012. *Government of Paper: The Materiality of Bureaucracy in Urban Pakistan.* Berkeley: University of California Press. https://doi.org/10.1525/california/9780520272149.001.0001.

Ialenti, Vincent. 2020. *Deep Time Reckoning: How Future Thinking Can Help Earth Now.* Cambridge, MA: MIT Press. http://dx.doi.org/10.7551/mitpress/12372.001.0001.

Ingold, Tim. 2000. *The Perception of the Environment: Essays on Livelihood, Dwelling & Skill.* London; New York: Routledge. https://doi.org/10.4324/9780203466025.

Jasanoff, Sheila. 1987. "Contested Boundaries in Policy-Relevant Science." *Social Studies of Science* 17 (2): 195–230. https://doi.org/10.2307/284949.

Jasanoff, Sheila. 2003. "Technologies of Humility: Citizen Participation in Governing Science." *Minerva* 41 (3): 223–44. https://doi.org/10.1023/A:1025557512320.

Jasanoff, Sheila. 2004. "Afterword." In *States of Knowledge: The Co-Production of Science and Social Order,* edited by Sheila Jasanoff, 272–82. London: Routledge.

Jasanoff, Sheila, and Hilton R. Simmet. 2017. "No Funeral Bells: Public Reason in a 'Post-Truth' Age." *Social Studies of Science* 47 (5): 751–70. https://doi.org/10.1177/0306312717731936.

Jensen, Casper Bruun. 2014. "Experiments in Good Faith and Hopefulness: Toward a Postcritical Social Science." *Common Knowledge* 20 (2): 337–62. https://doi.org/10.1215/0961754X-2422980.

Jensen, Casper Bruun, and Brit Ross Winthereik. 2013. *Monitoring Movements in Development Aid: Recursive Partnerships and Infrastructures.* Cambridge, MA: MIT Press. http://dx.doi.org/10.7551/mitpress/9301.001.0001.

Jiménez, Alberto Corsín. 2005. "After Trust." *Cambridge Anthropology* 25 (2): 64–78. https://www.jstor.org/stable/23820749.

Jiménez, Alberto Corsín. 2011. "Trust in Anthropology." *Anthropological Theory* 11 (2): 177–96. https://doi.org/10.1177/1463499611407392.

Kapferer, Jean-Noël. 2012. *The New Strategic Brand Management: Advanced Insights and Strategic Thinking.* London: Kogan Page Publishers.

Keane, Webb. 1994. "The Value of Words and the Meaning of Things in Eastern Indonesian Exchange." *Man* 29 (3): 605–29. https://doi.org/10.2307/2804345.

Keane, Webb. 1997. *Signs of Recognition: Powers and Hazards of Representation in an Indonesian Society.* Berkeley: University of California Press. http://dx.doi.org/10.1525/9780520917637.

Keller, Evelyn Fox. 1983. *A Feeling for the Organism: The Life and Work of Barbara McClintock.* San Francisco: W.H. Freeman.

Keller, Evelyn Fox. 1985. *Reflections on Gender and Science.* New Haven: Yale University Press. http://dx.doi.org/10.1119/1.15186.

Kipnis, Andrew B. 1997. *Producing Guanxi: Sentiment, Self, and Subculture in a North China Village.* Durham: Duke University Press.

Kipnis, Andrew B. 2008. "Audit Cultures: Neoliberal Governmentality, Socialist Legacy, or Technologies of Governing?" *American Ethnologist* 35 (2): 275–89. https://doi.org/10.1111/j.1548-1425.2008.00034.x.

Kohn, Eduardo. 2013. *How Forests Think: Toward an Anthropology beyond the Human.* Berkeley: University of California Press. http://dx.doi.org/10.1525/california/9780520276109.001.0001.

la paperson. 2017. *A Third University Is Possible.* Minneapolis: University of Minnesota Press. http://dx.doi.org/10.5749/9781452958460.

Laidlaw, James. 2000. "A Free Gift Makes No Friends." *The Journal of the Royal Anthropological Institute* 6 (4): 617–34. https://doi.org/10.1111/1467-9655.00036.

Lanzhe, dir. 2010. *Yak Dung*. Shan Shui Conservation Center and Nyanpo Yuzee Environmental Protection Association.

Lassiter, Luke E. 2005. *The Chicago Guide to Collaborative Ethnography*. Chicago: University of Chicago Press. http://dx.doi.org/10.7208/chicago/9780226467016.001.0001.

Latour, Bruno. 1987. *Science in Action: How to Follow Scientists and Engineers through Society*. Cambridge, MA: Harvard University Press.

Latour, Bruno. 1988. *The Pasteurization of France*. Translated by Alan Sheridan and John Law. Cambridge, MA: Harvard University Press.

Latour, Bruno. 2004a. *Politics of Nature: How to Bring the Sciences into Democracy*. Translated by Catherine Porter. Cambridge, MA: Harvard University Press. https://doi.org/10.4159/9780674039964.

Latour, Bruno. 2004b. "Why Has Critique Run Out of Steam? From Matters of Fact to Matters of Concern." *Critical Inquiry* 30 (2): 225–48. https://doi.org/10.1086/421123.

Lea, Tess. 2008. *Bureaucrats and Bleeding Hearts: Indigenous Health in Northern Australia*. Sydney: UNSW Press.

Leach, E. 1965. *Political Systems of Highland Burma: A Study of Kachin Social Structure*. Boston: Beacon Press.

Lei, Sean Hsiang-lin. 1999. "From Changshan to a New Anti-Malarial Drug: Re-Networking Chinese Drugs and Excluding Chinese Doctors." *Social Studies of Science* 29 (3): 323–58. https://doi.org/10.1177/030631299029003001.

Li, Ling. 2012. "The 'Production' of Corruption in China's Courts: Judicial Politics and Decision Making in a One-Party State." *Law & Social Inquiry* 37 (4): 848–77. https://doi.org/10.1111/j.1747-4469.2012.01285.x.

Li, Tania Murray. 2007. *The Will to Improve: Governmentality, Development, and the Practice of Politics*. Durham: Duke University Press. https://doi.org/10.1215/9780822389781.

Liboiron, Max. 2021. *Pollution Is Colonialism*. Durham: Duke University Press. https://doi.org/10.1215/9781478021445.

Liboiron, Max, Alex Zahara, and Ignace Schoot. 2018. "Community Peer Review: A Method to Bring Consent and Self-Determination into the Sciences." *Preprints* 2018060104. https://doi.org/10.20944/preprints201806.0104.v1.

Luhmann, Niklas. 2000. "Familiarity, Confidence, Trust: Problems and Alternatives." In *Trust: Making and Breaking Cooperative Relations*, edited by Diego Gambetta, 94–107. Oxford: Blackwell.

Mascarenhas, Michael. 2017. *New Humanitarianism and the Crisis of Charity: Good Intentions on the Road to Help*. Bloomington: Indiana University Press. http://dx.doi.org/10.2307/j.ctt200618x.

Mathews, Andrew S. 2011. *Instituting Nature: Authority, Expertise, and Power in Mexican Forests*. Cambridge, MA: MIT Press. http://dx.doi.org/10.7551/mitpress/9780262016520.001.0001.

Mathur, Nayanika. 2016. *Paper Tiger: Law, Bureaucracy and the Developmental State in Himalayan India*. Delhi: Cambridge University Press. http://dx.doi.org/10.1017/CBO9781316227367.

Maurer, Bill. 2005. *Mutual Life, Limited: Islamic Banking, Alternative Currencies, Lateral Reason*. Princeton: Princeton University Press. http://dx.doi.org/10.1515/9781400840717.

Mauss, Marcel. 2016. *The Gift*. Translated by Jane I. Guyer. Chicago: Hau Books.

Mavhunga, Clapperton Chakanetsa. 2018. *The Mobile Workshop: The Tsetse Fly and African Knowledge Production*. Cambridge, MA: MIT Press. http://dx.doi.org/10.7551/mitpress/10492.001.0001.

McLellan, Timothy. 2021a. "Impact, Theory of Change, and the Horizons of Scientific Practice." *Social Studies of Science* 51 (1): 100–120. https://doi.org/10.1177/0306312720950830.

McLellan, Timothy. 2021b. "Tools for an Efficient Witness: Deskilling Science and Devaluing Labor at an Agro-Environmental Research Institute." *HAU: Journal of Ethnographic Theory* 11 (2): 537–50. https://doi.org/10.1086/716421.

McLeod, Christopher, dir. 2013. *Standing on Sacred Ground*. Sacred Land Film Project.

Merry, Sally Engle. 2016. *The Seductions of Quantification: Measuring Human Rights, Gender Violence, and Sex Trafficking*. Chicago: Chicago University Press. http://dx.doi.org/10.7208/chicago/9780226261317.001.0001.

Mitchell, Timothy. 2002. *Rule of Experts: Egypt, Techno-Politics, Modernity*. Berkeley: University of California Press. http://dx.doi.org/10.1525/9780520928251.

Miyazaki, Hirokazu. 2003. "The Temporalities of the Market." *American Anthropologist* 105 (2): 255–65. https://doi.org/10.1525/aa.2003.105.2.255.

Miyazaki, Hirokazu. 2004. *The Method of Hope: Anthropology, Philosophy, and Fijian Knowledge*. Stanford: Stanford University Press. https://doi.org/10.1515/9781503624429.

Miyazaki, Hirokazu. 2013. *Arbitraging Japan: Dreams of Capitalism at the End of Finance*. Berkeley: University of California Press. http://dx.doi.org/10.1525/california/9780520273474.001.0001.

Miyazaki, Hirokazu. 2014. "Insistence and Response: On Ethnographic Replication." *Common Knowledge* 20 (3): 518–26. https://doi.org/10.1215/0961754X-2733063.

Moore, Sally Falk. 1978. *Law as Process: An Anthropological Approach*. London: Routledge and K. Paul.

Mosquera-Losada, M. R., J. J. Santiago-Freijanes, A. Pisanelli, M. Rois-Díaz, J. Smith, M. den Herder, G. Moreno, et al. 2018. "Agroforestry in the European Common Agricultural Policy." *Agroforestry Systems* 92 (4): 1117–27. https://doi.org/10.1007/s10457-018-0251-5.

Mosse, David. 2005. *Cultivating Development: An Ethnography of Aid Policy and Practice*. London: Pluto Press. http://dx.doi.org/10.2307/j.ctt18fs4st.

Mountz, Alison, Anne Bonds, Becky Mansfield, Jenna Loyd, Jennifer Hyndman, Margaret Walton Roberts, Ranu Basu, Risa Whitson, Roberta Hawkins, and Trina Hamilton. 2015. "For Slow Scholarship: A Feminist Politics of Resistance through Collective Action in the Neoliberal University." *ACME: An International Journal for Critical Geographies* 14 (4): 1235–59. https://acme-journal.org/index.php/acme/article/view/1058.

Munn, Nancy D. 1986. *The Fame of Gawa: A Symbolic Study of Value Transformation in a Massim (Papua New Guinea) Society*. Durham: Duke University Press.

Murphy, Michelle. 2021. "Reimagining Chemicals, with and against Technoscience." In *Reactivating Elements*, edited by Dimitris Papadopoulos, Maria Puig de la Bellacasa, and Natasha Myers, 257–79. Durham: Duke University Press. https://doi.org/10.1215/9781478021674-012.

Myers, Natasha. 2015. *Rendering Life Molecular: Models, Modelers, and Excitable Matter*. Durham: Duke University Press. https://doi.org/10.1215/9780822375630.

Myers, Natasha, and Joe Dumit. 2011. "Haptics: Haptic Creativity and the Mid-Embodiments of Experimental Life." In *A Companion to the Anthropology of the Body and Embodiment*, edited by Frances E. Mascia-Lees, 239–61. Oxford: Wiley-Blackwell. http://dx.doi.org/10.1002/9781444340488.ch13.

Nadasdy, Paul. 2007. "The Gift in the Animal: The Ontology of Hunting and Human–Animal Sociality." *American Ethnologist* 34 (1): 25–43. https://doi.org/10.1525/ae.2007.34.1.25.

Neal, Mark, and John Morgan. 2000. "The Professionalization of Everyone?: A Comparative Study of the Development of the Professions in the United Kingdom and Germany." *European Sociological Review* 16 (1): 9–26. https://doi.org/10.1093/esr/16.1.9.

Newfield, Christopher. 2003. *Ivy and Industry: Business and the Making of the American University, 1880–1980*. Durham: Duke University Press. https://doi.org/10.1215/9780822385202.

Newfield, Christopher. 2008. *Unmaking the Public University: The Forty-Year Assault on the Middle Class*. Cambridge, MA: Harvard University Press. https://doi.org/10.2307/j.ctv1cbn3np.

NSF (National Science Foundation). 2015. "IUSE / Professional Formation of Engineers: REvolutionizing Engineering and Computer Science Departments (RED) (Program Solicitation NSF 15-607)." NSF. https://www.nsf.gov/publications/pub_summ.jsp?ods_key=nsf15607.

NSF (National Science Foundation). 2022. "Improving Undergraduate STEM Education: Hispanic-Serving Institutions (HSI Program)—(Program Silicitation NSF 22-611)." NSF. https://www.nsf.gov/pubs/2022/nsf22611/nsf22611.htm?org=DUE#toc.

O'Reilly, Jessica. 2017. *The Technocratic Antarctic: An Ethnography of Scientific Expertise and Environmental Governance*. Ithaca: Cornell University Press. http://dx.doi.org/10.7591/9781501708367.

Osburg, John. 2013. *Anxious Wealth: Money and Morality among China's New Rich*. Stanford: Stanford University Press. http://dx.doi.org/10.1515/9780804785358.

Parreñas, Juno Salazar. 2018. *Decolonizing Extinction: The Work of Care in Orangutan Rehabilitation*. Durham: Duke University Press. https://doi.org/10.1215/9780822371946.

Pels, Peter. 2000. "The Trickster's Dilemma." In *Audit Cultures: Anthropological Studies in Accountability, Ethics and the Academy*, edited by Marilyn Strathern, 135–72. London: Routledge.

Philip, Kavita. 2004. *Civilizing Natures: Race, Resources, and Modernity in Colonial South India*. New Brunswick: Rutgers University Press.

Pia, Andrea E. 2016. "'We Follow Reason, Not the Law': Disavowing the Law in Rural China." *PoLAR: Political and Legal Anthropology Review* 39 (2): 276–93. https://doi.org/10.1111/plar.12194.

Pia, Andrea E. 2017. "Back on the Water Margin: The Ethical Fixes of Sustainable Water Provisions in Rural China." *Journal of the Royal Anthropological Institute* 23 (1): 120–36. https://doi.org/10.1111/1467-9655.12547.

Pickering, Andrew. 1995. *The Mangle of Practice: Time, Agency, and Science*. Chicago: University of Chicago Press. http://dx.doi.org/10.7208/chicago/9780226668253.001.0001.

Porter, Theodore M. 1995. *Trust in Numbers: The Pursuit of Objectivity in Science and Public Life*. Princeton: Princeton University Press. http://dx.doi.org/10.23943/princeton/9780691208411.001.0001.

Potter, Elizabeth. 2001. *Gender and Boyle's Law of Gases*. Bloomington: Indiana University Press.

Potter, Claire Bond. 2022. "The Hard Truths of the Academic-Labor Crisis: Even If Striking Workers Win, the System Is Still Rigged against Them." *The Chronicle of Higher Education*, 17 November 2022. https://www.chronicle.com/article/the-hard-truths-of-the-academic-labor-crisis.

Power, Michael. 1994. *The Audit Explosion*. London: Demos.

Power, Michael. 1997. *The Audit Society: Rituals of Verification*. Oxford: Oxford University Press. https://doi.org/10.1093/acprof:oso/9780198296034.001.0001.

Prescod-Weinstein, Chanda. 2022. *The Disordered Cosmos: A Journey into Dark Matter, Spacetime, and Dreams Deferred*. New York: Bold Type Books.

Puig de la Bellacasa, María. 2017. *Matters of Care: Speculative Ethics in More Than Human Worlds*. Minneapolis: University of Minnesota Press. https://www.jstor.org/stable /10.5749/j.ctt1mmfspt.

Rabeharisoa, Vololona, and Michel Callon. 2004. "Patients and Scientists in French Muscular Dystrophy Research." In *States of Knowledge: The Co-Production of Science and Social Order*, edited by Sheila Jasanoff, 142–60. London: Routledge.

Rabinow, Paul. 2011. *The Accompaniment: Assembling the Contemporary*. Chicago: University of Chicago Press. http://dx.doi.org/10.7208/chicago/9780226701714.001 .0001.

Raheja, Gloria Goodwin. 1988. *The Poison in the Gift: Ritual, Prestation, and the Dominant Caste in a North Indian Village*. Chicago: University of Chicago Press.

Raj, Kapil. 2007. *Relocating Modern Science: Circulation and the Construction of Knowledge in South Asia and Europe, 1650–1900*. Houndmills: Palgrave Macmillan. http://dx.doi.org/10.1057/9780230625310.

Rambo, Terry. 2007. "Observations on the Role of Improved Fallow Management in Swidden Agricultural Systems." In *Voices from the Forest: Integrating Indigenous Knowledge into Sustainable Upland Farming*, edited by Malcolm Cairns, 780–801. London: Routledge.

Rappaport, Joanne. 2005. *Intercultural Utopias Public Intellectuals, Cultural Experimentation, and Ethnic Pluralism in Colombia*. Durham: Duke University Press. https://doi.org/10.1215/9780822387435.

Readings, Bill. 1996. *The University in Ruins*. Cambridge, MA: Harvard University Press. http://dx.doi.org/10.2307/j.ctv1cbn3kn.

Reed, Adam. 2017. "An Office of Ethics: Meetings, Roles, and Moral Enthusiasm in Animal Protection." *Journal of the Royal Anthropological Institute*, 23 (S1): 166–181. https://doi.org/10.1111/1467-9655.12601.

Reed, Michael I. 1996. "Expert Power and Control in Late Modernity: An Empirical Review and Theoretical Synthesis." *Organization Studies* 17 (4): 573–97. https://doi .org/10.1177/017084069601700402.

Rothstein, Bo, and Dietlind Stolle. 2008. "The State and Social Capital: An Institutional Theory of Generalized Trust." *Comparative Politics* 40 (4): 441–59. https://doi.org /10.2307/20434095.

Rottenburg, Richard. 2009. *Far-Fetched Facts: A Parable of Development Aid*. Cambridge, MA: MIT Press. http://dx.doi.org/10.7551/mitpress/9780262182645.001.0001.

Rottenburg, Richard, Sally Engle Merry, Sung-Joon Park, and Johanna Mugler, eds. 2015. *The World of Indicators: The Making of Governmental Knowledge through Quantification*. Cambridge: Cambridge University Press. http://dx.doi.org/10.1017 /CBO9781316091265.

Rousseau, Denise M., Sim B. Sitkin, Ronald S. Burt, and Colin Camerer. 1998. "Not So Different after All: A Cross-Discipline View of Trust." *The Academy of Management Review* 23 (3): 393–404. https://doi.org/10.5465/amr.1998.926617.

Roy, Deboleena. 2008. "Asking Different Questions: Feminist Practices for the Natural Sciences." *Hypatia* 23 (4): 134–57. https://doi.org/10.1111/j.1527-2001.2008.tb01437.x.

Roy, Deboleena. 2018. *Molecular Feminisms: Biology, Becomings, and Life in the Lab*. Seattle: University of Washington Press.

Said, Edward W. 1994. *Representations of the Intellectual: The 1993 Reith Lectures*. New York: Pantheon Books.

Sanyang, Sidi, Sibiri Jean-Baptiste Taonda, Julienne Kuiseu, N'Tji Coulibaly, and Laban Konaté. 2016. "A Paradigm Shift in African Agricultural Research for Development: The Role of Innovation Platforms." *International Journal of Agricultural Sustainability* 14 (2): 187–213. https://doi.org/10.1080/14735903.2015.1070065.

Sauder, Michael, and Wendy Nelson Espeland. 2009. "The Discipline of Rankings: Tight Coupling and Organizational Change." *American Sociological Review* 74 (1): 63–82. https://doi.org/10.1177/000312240907400104.

Schinkel, Willem. 2016. "Making Climates Comparable: Comparison in Paleoclimatology." *Social Studies of Science* 46 (3): 374–95. https://doi.org/10.1177/0306312716633537.

Scott, James C. 1998. *Seeing like a State: How Certain Schemes to Improve the Human Condition Have Failed*. New Haven: Yale University Press. https://doi.org/10.12987/9780300252989.

Shapin, Steven, and Simon Schaffer. 1985. *Leviathan and the Air-Pump: Hobbes, Boyle, and the Experimental Life*. Princeton: Princeton University Press. https://doi.org/10.2307/j.ctt7sv46.

Shapiro, Judith. 2001. *Mao's War against Nature: Politics and the Environment in Revolutionary China*. Cambridge: Cambridge University Press. http://dx.doi.org/10.1017/CBO9780511512063.

Shore, Cris. 2008. "Audit Culture and Illiberal Governance Universities and the Politics of Accountability." *Anthropological Theory* 8 (3): 278–98. https://doi.org/10.1177/1463499608093815.

Shore, Cris, and Susan Wright. 1999. "Audit Culture and Anthropology: Neo-Liberalism in British Higher Education." *Journal of the Royal Anthropological Institute* 5 (4): 557–75. https://doi.org/10.2307/2661148.

Shore, Cris, and Susan Wright. 2015. "Audit Culture Revisited: Rankings, Ratings, and the Reassembling of Society." *Current Anthropology* 56 (3): 421–44. https://doi.org/10.1086/681534.

Sirman, Nükhet. 2016. "When Antigone Is a Man: Feminist 'Trouble' in the Late Colony." In *Vulnerability in Resistance*, edited by Judith Butler, Zeynep Gambetti, and Leticia Sabsay, 191–210. Durham: Duke University Press. https://doi.org/10.1215/9780822373490-010.

Sismondo, Sergio. 2017a. "Post-Truth?" *Social Studies of Science* 47 (1): 3–6. https://doi.org/10.1177/0306312717692076.

Sismondo, Sergio. 2017b. "Casting a Wider Net: A Reply to Collins, Evans and Weinel." *Social Studies of Science* 47 (4): 587–92. https://doi.org/10.1177/0306312717721410.

Sivaramakrishnan, Kalyanakrishnan. 1999. *Modern Forests: Statemaking and Environmental Change in Colonial Eastern India*. Stanford: Stanford University Press. http://dx.doi.org/10.1515/9781503617995.

Skallerup Bessette, Lee. 2019. "Contingency, Staff, Anxious Pedagogy—and Love." *Pedagogy* 19 (3): 525–29. https://doi.org/10.1215/15314200-7615502.

Skallerup Bessette, Lee. 2020a. "The Staff Are Not OK." *The Chronicle of Higher Education*, 30 October 2020. https://www.chronicle.com/article/the-staff-are-not-ok.

Skallerup Bessette, Lee. 2020b. "Staff Get Little to No Say in Campus Governance. That Must Change." *The Chronicle of Higher Education*, 22 September 2020. https://www.chronicle.com/article/staff-get-little-to-no-say-in-campus-governance-that-must-change.

Slawson, Nicola. 2017. "'Evidence Not Arrogance': UK Supporters Join Global March for Science." *The Guardian*, 22 April 2017. http://www.theguardian.com/science/2017/apr/22/evidence-not-arrogance-uk-supporters-join-global-march-for-science.

Smith, Trudi Lynn, Kate Hennessy, Stephanie Takaragawa, Fiona P. McDonald, and Craig Campbell. 2021. "Function and Form: The Ethnographic Terminalia Collective between Art and Anthropology." In *Collaborative Anthropology Today: A Collection of Exceptions*, edited by Boyer, Dominic, and George E. Marcus, 82–101. Ithaca: Cornell University Press. https://doi.org/10.7591/cornell/9781501753343.003.0006.

Smith-Morris, Carolyn. 2016. "When Numbers and Stories Collide: Randomized Control Trials and the Search for Ethnographic Fidelity in the Veterans Administration." In *Metrics: What Counts in Global Health*, edited by Vincanne Adams, 181–202. Durham: Duke University Press. https://doi.org/10.1215/9780822374480–009.

SOLR (Students Organizing for Labor Rights). 2022. "Petition for Northwestern & Compass to Respect Workers' Contract," 2022. https://docs.google.com/document/u/1/d/1wtsyiLFlWOEdouEbw6h3dRUDMSTLanAOMrXoplv-GNA/edit?usp=sharing&usp=embed_facebook.

Song, Priscilla. 2020. "Negotiating Evidence and Efficacy in Experimental Medicine." In *Can Science and Technology Save China?*, edited by Susan Greenhalgh and Li Zhang, 69–94. Ithaca: Cornell University Press. https://doi.org/10.7591/9781501747045–005.

Stein, Felix. 2018. "Anthropology's 'Impact': A Comment on Audit and the Unmeasurable Nature of Critique." *Journal of the Royal Anthropological Institute* 24 (1): 10–29. https://doi.org/10.1111/1467-9655.12749.

Stengers, Isabelle. 2011. "Comparison as a Matter of Concern." *Common Knowledge* 17 (1): 48–63. https://doi.org/10.1215/0961754X-2010–035.

Stengers, Isabelle. 2018. *Another Science Is Possible: A Manifesto for Slow Science*. Cambridge: Polity Press.

Stirling, Andy. 2005. "Opening Up or Closing Down? Analysis, Participation and Power in the Social Appraisal of Technology." In *Science and Citizens: Globalization and the Challenge of Engagement*, edited by Melissa Leach, Ian Scoones, and Brian Wynne, 218–31. London: Zed Books.

Strathern, Marilyn. 1988. *The Gender of the Gift: Problems with Women and Problems with Society in Melanesia*. Berkeley: University of California Press. http://dx.doi.org/10.1525/9780520910713.

Strathern, Marilyn. 2000. "The Tyranny of Transparency." *British Educational Research Journal; Oxford* 26 (3): 309. https://doi.org/10.1080/713651562.

Strathern, Marilyn. 2006. "A Community of Critics? Thoughts on New Knowledge." *Journal of the Royal Anthropological Institute* 12 (1): 191–209. https://doi.org/10.1111/j.1467-9655.2006.00287.x.

Subramaniam, Banu, and Angela Willey. 2017a. "Introduction: Feminism's Sciences." *Catalyst: Feminism, Theory, Technoscience* 3 (1): 1–23. https://doi.org/10.28968/cftt.v3i1.28784.

Subramaniam, Banu, and Angela Willey. 2017b. "Introduction: Remaking Science(s)." *Catalyst: Feminism, Theory, Technoscience* 3 (2): 1–9. https://doi.org/10.28968/cftt.v3i2.28839.

Taylor, Janelle S. 2003. "Confronting 'Culture' in Medicine's 'Culture of No Culture.'" *Academic Medicine* 78 (6): 555–59. https://doi.org/10.1097/00001888-200306000–00003.

Thornton, Margaret. 2014. *Through a Glass Darkly: The Social Sciences Look at the Neoliberal University*. Canberra: ANU Press.

Tracy, Megan. 2016. "Multimodality, Transparency, and Food Safety in China." *PoLAR: Political and Legal Anthropology Review* 39 (S1): 34–53. https://doi.org/10.1111/plar.12170.

Tsing, Anna Lowenhaupt. 2005. *Friction: An Ethnography of Global Connection*. Princeton: Princeton University Press. https://doi.org/10.2307/j.ctt7s1xk.

Tsing, Anna Lowenhaupt. 2015. *The Mushroom at the End of the World: On the Possibility of Life in Capitalist Ruins*. Princeton: Princeton University Press. https://doi.org/10.2307/j.ctvc77bcc.

Vaughn, Sarah E. 2017. "Disappearing Mangroves: The Epistemic Politics of Climate Adaptation in Guyana." *Cultural Anthropology* 32 (2): 242–68. https://doi.org/10.14506/ca32.2.07.

Vaughn, Sarah E. 2017. 2022. *Engineering Vulnerability: In Pursuit of Climate Adaptation*. Durham: Duke University Press. https://doi.org/10.1215/9781478022725.

Venkatesan, Soumhya. 2011. "The Social Life of a 'Free' Gift." *American Ethnologist* 38 (1): 47–57. https://doi.org/10.1111/j.1548-1425.2010.01291.x.

Vonnegut, Kurt. 1963. *Cat's Cradle*. New York: Laurel.

Vrieze, Jop de. 2017. "'Science Wars' Veteran Has a New Mission." *Science* 358 (6360): 159–159. https://doi.org/10.1126/science.358.6360.159.

Wagner, Roy. 1986. *Symbols That Stand for Themselves*. Chicago: University of Chicago Press.

Watanabe, Chika. 2015. "Commitments of Debt: Temporality and the Meanings of Aid Work in a Japanese NGO in Myanmar." *American Anthropologist* 117 (3): 468–79. https://doi.org/10.1111/aman.12287.

Watanabe, Chika. 2019. *Becoming One: Religion, Development, and Environmentalism in a Japanese NGO in Myanmar*. Honolulu: University of Hawai'i Press. https://doi.org/10.2307/j.ctv7r434x.

Webb, Martin. 2019. "Seeking Signs of Transparency: Audit, Materiality, and Monuments to Active Citizenship in New Delhi." *Journal of the Royal Anthropological Institute* 25 (4): 698–720. https://doi.org/10.1111/1467-9655.13129.

Weber, Max. 1946. "Science as a Vocation." In *From Max Weber: Essays in Sociology*, edited by Hans Gerth and C. Wright Mills. New York: Oxford University Press.

Weiner, Annette B. 1976. *Women of Value, Men of Renown: New Perspectives in Trobriand Exchange*. Austin: University of Texas Press.

Welker, Marina. 2014. *Enacting the Corporation: An American Mining Firm in Post-Authoritarian Indonesia*. Berkeley: University of California Press. http://dx.doi.org/10.1525/california/9780520282308.001.0001.

White, Sarah C. 2006. "The 'Gender Lens': A Racial Blinder?" *Progress in Development Studies* 6 (1): 55–67. https://doi.org/10.1191/1464993406ps127oa.

Wilensky, Harold L. 1964. "The Professionalization of Everyone?" *American Journal of Sociology* 70 (2): 137–58. https://doi.org/10.1086/223790.

Wright, Susan, and Davydd J. Greenwood. 2017. "Universities Run for, by, and with the Faculty, Students and Staff: Alternatives to the Neoliberal Destruction of Higher Education." *Learning and Teaching* 10 (1): 42–65. https://doi.org/10.3167/latiss.2017.100104.

Wynne, Brian. 1992. "Misunderstood Misunderstanding: Social Identities and Public Uptake of Science." *Public Understanding of Science* 1 (3): 281–304. https://doi.org/10.1088/0963-6625/1/3/004.

Yan, Yunxiang. 1996. *The Flow of Gifts: Reciprocity and Social Networks in a Chinese Village*. Stanford: Stanford University Press.

Yang, Mayfair Mei-hui. 1994. *Gifts, Favors, and Banquets: The Art of Social Relationships in China*. Ithaca: Cornell University Press. http://dx.doi.org/10.7591/9781501713057.

Yang, Qing, and Wenfang Tang. 2010. "Exploring the Sources of Institutional Trust in China: Culture, Mobilization, or Performance?" *Asian Politics & Policy* 2 (3): 415–36. https://doi.org/10.1111/j.1943-0787.2010.01201.x.

Yarrow, Thomas. 2019. *Architects: Portraits of a Practice*. Ithaca: Cornell University Press. http://dx.doi.org/10.7591/cornell/9781501738494.001.0001.

Yarrow, Thomas, and Soumhya Venkatesan. 2012. "Anthropology and Development: Critical Framings." In *Differentiating Development: Beyond an Anthropology of Critique*, edited by Soumhya Venkatesan and Thomas Yarrow, 1–22. New York: Berghahn Books.

Zhan, Mei. 2001. "Does It Take a Miracle? Negotiating Knowledges, Identities, and Communities of Traditional Chinese Medicine." *Cultural Anthropology* 16 (4): 453–80. https://doi.org/10.1525/can.2001.16.4.453.

Zhang, Li. 2001. *Strangers in the City: Reconfigurations of Space, Power, and Social networks within China's Floating Population.* Stanford: Stanford University Press.

Zhang, Shaoying, and Derek McGhee. 2017. *China's Ethical Revolution and Regaining Legitimacy: Reforming the Communist Party through Its Public Servants.* Cham: Palgrave Macmillan. http://dx.doi.org/10.1007/978-3-319-51496-3.

Zhao, Shukai. 2007. "The Accountability System of Township Governments." Translated by Tes Wang. *Chinese Sociology & Anthropology* 39 (2): 64–73. https://doi.org/10.2753/CSA0009-4625390207.

Index

www.ingramcontent.com/pod-product-compliance
Lightning Source LLC
Chambersburg PA
CBHW030850270326
41928CB00008B/1301